BIOZONE

AQA
BIOLOGY 2
A-Level Year 2/AL

MODEL ANSWERS

This model answer booklet is a companion publication to provide answers for the activities in the AQA Biology 2 Student Workbook. These answers have been produced as a separate publication to keep the cost of the workbook itself to a minimum. All answers to set questions are provided, but chapter reviews are the student's own and no model answer is set. Working and explanatory notes have been provided where clarification of the answer is appropriate.

ISBN 978-1-927309-22-3

Copyright © 2016 Richard Allan
Published by BIOZONE International Ltd

PHOTOCOPYING PROHIBITED

including photocopying under a photocopy licence scheme such as CLA

Additional copies of this Model Answers book may be purchased directly from the publisher.

BIOZONE Learning Media (UK) Ltd.

Telephone local:	01283 530 366
Telephone international:	+44 1283 530 366
Fax local:	01283 831 900
Fax international:	+44 1283 831 900
Email:	sales@biozone.co.uk

www.**BIOZONE**.co.uk

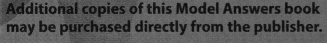

CONTENTS AQA BIOLOGY 2

Energy transfers in and between organisms

Organisms respond to changes in their environment

CONTENTS AQA BIOLOGY 2

Genetics, populations, evolution and ecosystems

CONTENTS AQA BIOLOGY 2

The control of gene expression

Note: Erratum (chapter 1, introduction)
#30 Describe the cycling of nitrogen and **phosphorus** between the abiotic and biotic environments, including reference to …
• **saprobionts** in decomposition and ammonification
• bacteria in … **denitrification**

1. Energy in Cells (page 3)
1. (a) Photosynthesis: Carbon dioxide and water.
   (b) Cellular respiration: Oxygen and glucose.

2. Glucose (or pyruvate).

3. Solar energy (the Sun).

4. Food (plants and other animals).

2. Photosynthesis (page 4)
1. (a) and (b) in any order:
   (a) Grana: Stacks of thylakoid membranes containing chlorophyll molecules. They are the site of the light dependent reactions of photosynthesis, which involve light energy capture via photosystems I and II.
   (b) Stroma: The liquid interior of the chloroplast in which the light independent phase takes place. This biochemical process involves carbon fixation (production of carbohydrate) via the Calvin cycle.

2. (a) Carbon dioxide: Comes from the air (through stomata) and provides carbon and oxygen for the production of glucose. Some oxygen molecules contribute to producing H_2O.
   (b) Oxygen: Comes from CO_2 gas (through stomata) and H_2O (via the roots and vascular system). The oxygen from the CO_2 is incorporated into glucose and H_2O. The oxygen from water is given off as free oxygen (a waste product).
   (c) Hydrogen: Comes from water (via the roots and vascular system) from the soil. This hydrogen is incorporated into glucose and H_2O. Note: To clarify: isotope studies show that the carbon and oxygen in the carbohydrate comes from CO_2, while the free oxygen comes from H_2O.

3. Glucose is used as the fuel for cellular respiration, or used to construct cellulose, starch, or disaccharide molecules (e.g. sucrose). Oxygen is required for aerobic respiration and water is recycled and even reused for photosynthesis.

3. Chloroplasts (page 5)
1. (a) Stroma   (d) Granum
   (b) Stroma lamellae   (e) Thylakoid
   (c) Outer membrane   (f) Inner membrane

2. (a) Chlorophyll is found in the thylakoid membranes.
   (b) Chlorophyll is a membrane-bound pigment found in and around the photosystems embedded in the membranes. Light capture by chlorophyll is linked to electron transport in the light dependent reactions.

3. The internal membranes provide a large surface area for binding chlorophyll molecules and capturing light. Membranes are stacked in such a way that they do not shade each other.

4. Chlorophyll absorbs blue and red light but reflects green light, so leaves look green to the human eye.

4. Pigments and Light Absorption (page 6)
1. The absorption spectrum of a pigment is that wavelength of the light spectrum absorbed by a pigment, e.g. chlorophyll absorbs red and blue light and appears green. Represented graphically, the absorption spectrum shows the relative amounts of light absorbed at different wavelengths.

2. Accessory pigments absorb light wavelengths that chlorophyll a cannot absorb, and they pass their energy on to chlorophyll a. This broadens the action spectrum over which chlorophyll a can fuel photosynthesis.

5. Separation of Pigments by Chromatography (page 7)
1. (a) and (b)
   A: Rf value 0.92. Pigment: carotene
   B: Rf value 0.53. Pigment: Chlorophyll a
   C: Rf value: 0.46. Pigment: Chorophyll b
   *Note: exact replication of the Rf values may not always

occur. In this example, Chlorophyll a and b can be recognised due to their relative positions and closeness to the correct Rf values.
   D: Rf value: 0.30. Pigment: Xanthophyll

2. There should be no effect on the Rf values but the amount of separation will be reduced.

6. Light Dependent Reactions (page 8)
1. NADP: Carries H_2 from the light dependent phase to the light independent reactions.

2. Chlorophyll molecules trap light energy and produce high energy electrons. These are used to make ATP and NADPH. The chlorophyll molecules also split water, releasing H^+ for use in the light independent reactions and liberating free O_2.

3. Light dependent (D) phase takes place in the grana (thylakoid membranes) of the chloroplast and requires light energy to proceed. The light dependent phase generates ATP and reducing power in the form of NADPH. The electrons and hydrogen ions come from the splitting of water.

4. The ATP synthesis is coupled to electron transport. When the light strikes the chlorophyll molecules, high energy electrons are released by the chlorophyll molecules. The energy lost when the electrons are passed through a series of electron carriers is used to generate ATP from ADP and phosphate.

   Note: ATP is generated (in photosynthesis and cellular respiration) by chemiosmosis. As the electron carriers pick up the electrons, protons (H^+) pass into the space inside the thylakoid, creating a high concentration of protons there. The protons return across the thylakoid membrane down a concentration gradient via the enzyme complex, ATP synthase, which synthesises ATP.

5. (a) Non-cyclic phosphorylation: Generation of ATP using light energy during photosynthesis. The electrons lost during this process are replaced by the splitting of water.
   (b) The term non-cyclic photophosphorylation is also (commonly) used because it indicates that the energy for the phosphorylation is coming from light.

6. (a) In cyclic photophosphorylation, the electrons lost from photosystem II are replaced by those from photosystem I rather than from the splitting of water. ATP is generated in this process, but not NADPH. Note: In the cell, both cyclic and non-cyclic photophosphorylation operate to different degrees to balance production of NADPH and ATP.
   (b) The non-cyclic path produces ATP and NADH in roughly equal quantities but the Calvin cycle uses more ATP than NADPH. The cyclic pathway of electron flow makes up the difference.

7.

	Non-cyclic phosphorylation	Cyclic phosphorylation
Photosystem	I and II	I
Energy carrier produced	NADPH, ATP	ATP
Photolysis of water	Yes	No
Production of oxygen	Yes	No

7. Light Independent Reactions (page 10)
1. (a) 6   (d) 12   (g) 2
   (b) 6   (e) 12   (h) 1
   (c) 12   (f) 6

2. RuBisCo catalyses the reaction that splits CO_2 and joins it with ribulose 1,5-bisphosphate. It fixes carbon from the atmosphere.

3. Triose phosphate (note that you may also see this referred to as glyceraldehyde-3-phosphate, GALP, G3P or PGAL)

4. $6CO_2 + 18ATP + 12 NADPH + 12H^+$
   $\rightarrow 1$ glucose $+18ADP + 18Pi +12 NADP+ + 6H_2O$

5. The Calvin cycle will cease in the dark in most plants because the light dependent reactions stop, therefore no NADPH or ATP is produced. At night, stomata also close, reducing levels of CO_2 (there will still be some CO_2 in the leaf as a waste product of respiration).

8. Experimental Evidence for Photosynthesis (page 11)
1. $DCPIP + 2H^+ + 2e^- \rightarrow DCPIPH_2$

2. Hill's experiment showed that water must be the source of the oxygen liberated in photosynthesis.

3. The reactions happen very quickly. By taking the sample only seconds apart, each step of the reaction can be worked out by recording the order of appearance of the reaction products.

9. Investigating Enzymes in Photosynthesis (page 12)
1. If light intensity affects the rate of dehydrogenase activity, then DCPIP will fade at a faster rate when exposed to a high light intensity than when exposed to a lower light intensity.

2.

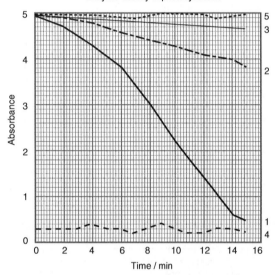

Enzyme activity in photosynthesis

3. (a) Tube 4 is a control to test if the DCPIP is responsible for the colour change.
   (b) Tube 5 is a control to test that the colour change is due to the leaf extract (not an unrelated reaction or light alone).

4. Tube 3 can only be exposed to the light for a short time to test that DCPIP does not fade in the dark (no reactions occurring).

5. The variation is due to natural variation of the absorbance reading (e.g minor variations in angle of tube etc).

6. Dehydrogenase activity (as an indirect measure of the rate of the light dependent reactions of photosynthesis) increases with an increase in light intensity.

10. The Fate of Triose Phosphate (page 13)
1. Two triose phosphate molecules are used to form a glucose molecule.

2. Glucose has three main fates: storage, building macromolecules, or production of usable energy (ATP).

3. In plants, glucose can be converted to (any one of): fructan (energy storage in vacuoles), starch (energy storage in plastids), or cellulose (cell wall component).

4. Glucose is produced using a ^{13}C or ^{14}C. When the glucose is used to make other molecules, the C isotopes can be detected by the radioactivity or by density change of tissues or products into which the carbon has been incorporated.

11. Factors Affecting Photosynthesis (page 14)
1. (a) CO_2 concentration: Photosynthetic rate increases as the CO_2 concentration increases, and then levels off.
   (b) Light intensity: Photosynthetic rate increases rapidly as

the light intensity increases and then levels off.
   (c) Temperature: Increased temperature increases the photosynthetic rate, but this effect is not marked at low concentrations of CO_2.

2. The photosynthetic rate is determined by the rate at which CO_2 enters the leaf. When this declines because of low atmospheric levels, so too does photosynthetic rate.

3. Plants close their stomata during hot, windy conditions to reduce water loss thus reducing there CO_2 uptake and also reducing photosynthetic rates.

12. Overcoming Limiting Factors in Photosynthesis (page 15)
1. CO_2 enrichment increases the rate of photosynthesis and so also the formation of dry plant matter (growth) and total yield (flowers or fruit).

2. The two primary economic considerations include the capital cost of equipment and the cost of operation (e.g. powering the equipment). If production costs exceed operational costs, the controlled environment will be viable.

3. Temperature, carbon dioxide concentration, humidity, soil composition, air movement, and light intensity are some of the abiotic factors controlled in a greenhouse environment.

13. Investigating Photosynthetic Rate (page 16)
1. Missing figures for bubbles per minute are (in order of low light intensity to high light intensity): 0, 2, 3, 4, 6, 11, 11.67

2.

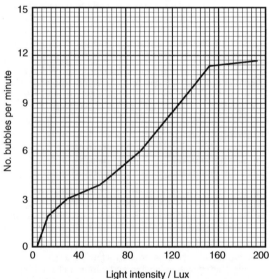

Bubbles produced by *Cabomba*

3. Photosynthetic rate increases with increasing light intensity. Although light intensity theoretically drops off at a constant rate, this may not practically occur due to shadows, variations in the equipment being used, or other light pollution. It is therefore better to measure the light intensity rather than infer light intensity from distance.

4. The gas was oxygen.

5. Instead of counting the bubbles (which could vary in volume) the gas could be collected and the volume produced measured. This could be done by displacement of water in a graduated cylinder or by using a photosynthometer.

14. The Role of ATP in Cells (page 17)
1. Organisms need to respire so that the energy in food can be converted, via a series of reactions, into the energy yielding molecule, ATP, which powers metabolic reactions.

2. (a) Mitochondria are the site for the Krebs cycle and ETS stages of cell respiration and ATP production.
   (b) The mitochondrion is separated into regions

(compartmentalised) by membranes. This allows certain metabolic reactions, together with their required enzymes, to be isolated in a specific region. All of the required components are in one place, increasing efficiency.

3. Maintaining body temperature (thermoregulation) requires energy input. ATP is required for the muscular activity involved in shivering (used to heat the body). ATP is also required for the secretion of sweat (used to cool the body).

15. ATP and Energy (page 18)
1. (a) The hydrolysis of ATP is coupled to the formation of a reactive intermediate, which can do work. Effectively, the hydrolysis of ATP to ADP + Pi releases energy.
   (b) Like a rechargeable battery, the ADP/ATP system alternates between high energy and low energy states. The addition of a phosphate to ADP recharges the molecule so that it can be used for cellular work.

2. Glucose

3. The folded inner membrane of a mitochondrion greatly increases the surface area. This allows more ATP synthase molecules to be accommodated on the membrane and therefore increases the ability to produce ATP.

4. Highly active cells require a lot of energy (ATP) to move. Therefore, they have large numbers of mitochondria so that enough ATP can be produced to meet their energy demands.

16. ATP Production in Cells (page 19)
1. (a) Glycolysis: cytoplasm
   (b) Link reaction: matrix of mitochondria
   (c) Krebs cycle: matrix of mitochondria
   (d) Electron transport chain: cristae (inner membrane surface) of mitochondria.

2. The ATP generated in glycolysis and the Krebs cycle is generated by substrate level phosphorylation, i.e. transfer of a phosphate group directly from a substrate to ADP. In contrast, the ATP generated via the electron transport chain is through oxidative phosphorylation, a step-wise series of redox reactions that provide the energy for forming ATP. Oxidative phosphorylation yields much more ATP per glucose than substrate level phosphorylation.

17. The Biochemistry of Respiration (page 20)
1.

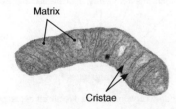

Matrix

Cristae

2. The link reaction prepares pyruvate for entry in the Krebs cycle. Carbon dioxide is removed and coenzyme A is added.

3. (a) 6   (b) 3   (c) 2   (d) 6   (e) 5   (f) 4

4. (a) Glycolysis: 2 ATPs
   (b) Krebs cycle: 2 ATPs
   (c) Electron transport chain: 34 ATPs
   (d) Total produced: 38 ATPs

5. The carbon atoms lost are lost as CO_2 molecules.

6. During oxidative phosphorylation, ADP is phosphorylated to ATP. Electrons passed along the electron transport chain are used to pump hydrogen ions across the inner membrane of the mitochondria. The flow of hydrogen ions back across the membrane is coupled to the phosphorylation of ADP to ATP. Oxygen is the final electron acceptor, reducing hydrogen to water. Because oxygen is the final acceptor the process is called oxidative phosphorylation.

18. Chemiosmosis (page 22)
1. In chemiosmosis, ATP synthesis is coupled to electron transport and movement of hydrogen ions. Energy from the passage of electrons along the chain of electron carriers is used to pump protons (H^+), against their concentration gradient, into the intermembrane space, creating a high concentration of protons there. The protons return across the membrane down a concentration gradient via the enzyme complex, ATP synthase, which synthesises the ATP.

2. Elevating the H^+ concentration outside the exposed inner mitochondrial membranes would cause them to move down their concentration gradient via ATP synthase, generating ATP.

3. A suspension of isolated chloroplasts would become alkaline because protons would be removed from the medium as ATP was generated.

4. (a) By placing chloroplasts in an acid medium, the thylakoid interior was acidified. Transfer to an alkaline medium established a proton gradient from the thylakoid interior to the medium.
   (b) The protons could flow down the concentration gradient established, via ATP synthase, and generate ATP.

19. Anaerobic Pathways (page 23)
1. **Aerobic respiration** requires the presence of oxygen and produces a lot of useable energy (ATP). **Fermentation** does not require oxygen and uses an alternative H^+ acceptor. There is little useable energy produced (the only ATP generated is via glycolysis).

2. (a) 2 ÷ 38 x 100 = 5.3% efficiency
   (b) Only a small amount of the energy of a glucose molecule is released in anaerobic respiration. The remainder stays locked up in the molecule.

3. The build up of ethanol, which is toxic to cells, inhibits further metabolic activity.

20. Investigating Fermentation in Yeast (page 24)
1. $C_6 H_{12} O_6 \rightarrow C_2 H_5 OH + 2CO_2$

2. Calculated rate of CO_2 production, group 1:
   (a) None: 0 cm^3min^{-1} or 0 cm^3s^{-1}
   (b) Glucose: 0.443 cm^3min^{-1} or 7.4 x 10^{-3} cm^3s^{-1}
   (c) Maltose: 0.24 cm^3min^{-1} or 4.0 x 10^{-3} cm^3s^{-1}
   (d) Sucrose: 0.191 cm^3min^{-1} or 3.2 x 10^{-3} cm^3s^{-1}
   (e) Lactose: 0 cm^3min^{-1} or 0 cm^3s^{-1}

3. The assumptions made are that 24°C and pH 4.5 provide suitable (even optimal) conditions for yeast fermentation. This is reasonable as it has been stated in the background that the literature cites warm, slightly acidic conditions as optimal for baker's yeast.

4. Graph of results: see top of the next page.

5. (a) Time
   (b) 0 - 95 minutes in 5 minute increments.
   (c) Minutes

6. (a) Volume of CO_2 produced
   (b) cm^3

7. (a) The fermentation rates were greatest for the substrate glucose, with a CO_2 yield approximately twice that for maltose and sucrose. Maltose and sucrose were similar to each other, while there was no fermentation of lactose.
   (b) CO_2 production is highest when glucose is the substrate because it is directly available as a fuel and requires no initial hydrolysis to use.
   (c) Maltose (glucose-glucose) and sucrose (glucose-fructose) must first be hydrolysed before the glucose is available (the fructose from sucrose must also be isomerised to glucose).
   (d) Lactose cannot be metabolised by yeast, presumably because yeast lack the enzyme to hydrolyse the galactose and glucose that form this disaccharide.

8. CO_2 production would increase more rapidly.

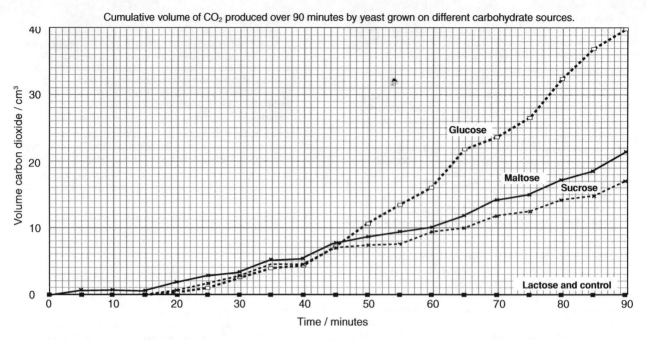

Cumulative volume of CO_2 produced over 90 minutes by yeast grown on different carbohydrate sources.

21. Investigating Aerobic Respiration in Yeast (page 26)

1.

Time / hr	Mean absorbance		
	25 °C	40 °C	55 °C
1.5	1.12	2.32	4.11
3.0	1.96	5.85	8.86
4.5	2.75	7.84	9.79

2.

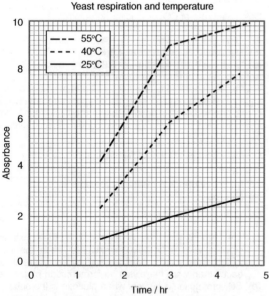

3. The yeast multiplied changing the clarity of the solution.

4. As temperature increases, respiration rate also increases.

22. Energy in Ecosystems: Food Chains (page 27)

1. (a) The sun.
   (b) The energy is converted to biomass through the process of photosynthesis.
   (c) Refer to the diagram, below

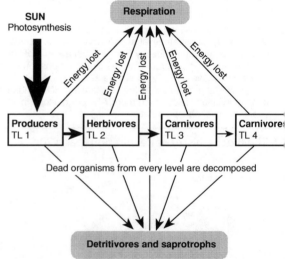

2. Energy is transferred in the chemical bonds in biomass.

3. (a) Producers obtain energy from the sun via photosynthesis the process by which the energy is converted to biomass
   (b) Consumers obtain energy by eating other organisms.
   (c) Detritivores obtain energy from eating dead organic matter.
   (d) Saprotrophs obtain energy by extracellular digestion of dead material.

23. What is Primary Productivity? (page 28)

1. Estuaries, swamps, and tropical rainforests typically have higher temperatures (than open water or temperate forest), ample water, and readily available nitrogen thus yielding higher net productivities.

2. (a) [initial - dark] = 8 mg O_2 L^{-1} - 5 mg O_2 L^{-1} = 3 mg O_2 L^{-1}
   (b) [light - initial] = 10 mg O_2 L^{-1} - 8 mg O_2 L^{-1} = 2 mg O_2 L^{-1}
   (c) [light - dark] = 10 mg O_2 L^{-1} - 5 mg O_2 L^{-1} = 5 mg O_2 L^{-1}
   (or NPP + R = 2 mg O_2 L^{-1} + 3 mg O_2 L^{-1} = 5 mg O_2 L^{-1})
   (d) 9.4 grams of glucose ÷ 2 = 4.7 grams of glucose

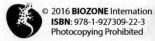

© 2016 **BIOZONE** Internation
ISBN: 978-1-927309-22-3
Photocopying Prohibited

24. Productivity and Trophic Efficiency (page 29)

1. (a)- (c) and any of the following:
   - Amount and availability of light for photosynthesis. This is higher in the tropics.
   - Temperature. Higher temperatures are generally associated with higher productivity.
   - Availability of water. Photosynthesis (and therefore productivity) will be limited when water is scarce.
   - Availability of nutrients. Nutrient limitations will limit plant growth and lower productivity.

2. (a) Secondary productivity is the rate of production of consumer biomass.
   (b) High palatability and turnover of biomass contribute to high secondary productivity. The large number of trophic connections also reduce energy losses within the system.

25. Energy Inputs and Outputs (page 30)

1. (a) Metabolic wastes as urine, faeces, and carbon dioxide.
   (b) Heat as a result of cellular metabolism

2. (a) Corn: 138.0 ÷ 8548 X 100 = 1.61 %
       Pasture: 24.4 ÷ 1971 x 100 = 1.24 %
   (b) Corn: 32.2 ÷ 138.0 x 100 = 23.3%
       Pasture: 3.7 ÷ 24.4 x 100 = 15.2%
   (c) Corn: 105 ÷ 138 x 100 = 76.7%
       Pasture: 20.7 ÷ 24.4 x 100 = 84.8%
   (d) The mature pasture

3. If N = 20% of I (700 kJ), then N = 0.2 x 700 = 140.
   Energy lost as F and R = 700 − 140 = 560 kJ.

4. See percent biomass column in table below

5. See Energy columns in table below

6. See NPP column in table below

Age in days	Percent biomass	Energy in 10 plants / kJ	Energy per plant / kJ	NPP / kJ plant^{-1} d^{-1}
7	21.4	76.4	7.6	1.09
14	24.2	169.3	16.9	1.2
21	28.1	282.1	28.2	1.34

7. Transfer NPP figures above to the first column of Table 2. Mean calculation below:

Age in days	Mean NPP / kJ plant^{-1} d^{-1}
7	1.09
14	1.21
21	1.33

8.

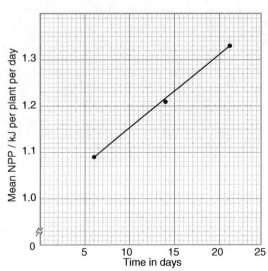

9. (a) NPP increases over time in a linear fashion.
   (b) NPP increases over time because anabolic processes in the plants increase the amount of plant tissue that can be used both to capture more of the sun's energy and as energy for the next trophic level (e.g. more sugars, more cellulose, more chlorophyll).

10. To determine the GPP of *B. rapa* you would need to know how much energy the plants have lost in respiration.

11. A basic methodology as follows:

    Introduction:

    Of the NPP from the brussels sprouts that is consumed by the caterpillars, some will be used in cellular respiration, some will be available to secondary consumers (the net 2° production) and some will be lost as waste products (frass).

    Note that in the energy values provided in the activity, plant mass contains different percentages of organic compounds to animal material and frass. Different organic compounds contain different amounts of energy per gram.

    Method

    - Put a small amount of brussels sprouts in an aerated container with 10, 12 day old caterpillars. Weigh the brussels sprouts and the caterpillars to obtain a wet mass.

    - After 3 days, disassemble the container and record the mass of the remaining brussels sprouts, caterpillars, and frass.

    - In separate containers, dry the remaining brussels sprouts, the caterpillars, and the frass in a drying oven to obtain the dry mass.

    - For brussels sprouts tabulate:
      (a) Wet mass on days 1 and 3
      (b) Dry mass on day 3 (after drying)

    - On the table for brussels sprouts calculate:
      (c) Percent biomass on day 3 (dry÷wet x 100). Assume percent biomass on day 1 is the same.
      (d) Plant energy on days 1 and 3 (wet mass x percent biomass x 18.2 kJ)
      (e) Plant energy eaten per caterpillar in 3 days (plant energy day 1 − plant energy day 3 ÷ 10 caterpillars)

    Assumptions: That the percent biomass of brussels sprouts (and caterpillars) on day 1 is the same as the calculated value from day 3. You cannot calculate percent biomass on day 1 because it would mean destroying the food and the caterpillars.

    - For caterpillars tabulate:
      (f) Wet mass on days 1 and 3 and mass gained (g)
      (g) Wet mass per individual on days 1 and 3 and mass gained (g) (mass gained ÷ 10)
      (h) Dry mass on day 3 (after drying)

    - On the table for caterpillars calculate:
      (i) Caterpillar percent biomass on day 3 (dry÷wet x 100). Assume percent biomass on day 1 is the same.
      (j) **Net secondary production**
          Energy production per caterpillar on days 1 and 3 (individual wet mass x percent biomass x 23.0 kJ).
          Net secondary production = kJ gained per caterpillar = Energy production per caterpillar day 3 − day 1
      (k) **Efficiency of energy transfer, producers to consumers**:
          Energy production per caterpillar day 3 − energy production per caterpillar day 1 (j) ÷ plant energy eaten per caterpillar (e) x 100 should be ~10% (or less)

    - For frass tabulate:
      (l) Dry mass of frass from 10 caterpillars

    - On the table for frass calculate:
      (m) Frass energy = frass dry mass x 19.87 kJ
      (n) Energy of frass from 1 caterpillar (frass energy ÷ 10)

    To calculate respiratory losses:
    Plant energy eaten per caterpillar (e)
    minus Energy production per caterpillar (net 2° production) (j)
    minus Frass energy per caterpillar (n)

26. Food Webs (page 32)
1. (a) Carnivore   (d) Autotroph
   (b) Detritivore   (e) Herbivore (when young)
   (c) Detritus

2. Most energy is lost from the system as heat, so very little is transferred to the next level. After six links there is very little energy left in the system (not enough energy available to support the organisms in another level).

27. Energy Budget in an Ecosystem (page 33)
1. (a) 14 000   (b) 180
   (c) 35   (d) 100

2. Solar energy

3. A. Photosynthesis
   B. Eating/feeding/ingestion
   C. Respiration
   D. Export (lost from this ecosystem to another)
   E. Decomposers and detritivores feeding on other decomposers and detritivores
   F. Radiation of heat to the atmosphere
   G. Excretion/egestion/death

4. (a) $1\ 700\ 000 \div 7\ 000\ 000 \times 100 = 24.28\%$
   (b) It is reflected. Plants appear green because those wavelengths are not absorbed. Reflected light falls on other objects as well as back into space.

5. (a) $87\ 400 \div 1\ 700\ 000 \times 100 = 5.14\%$
   (b) $1\ 700\ 000 - 87\ 400 = 1\ 612\ 600\ (94.86\%)$
   (c) Most of the energy absorbed by producers is not used in photosynthesis. This excess energy, which is not fixed, is lost as heat (although the heat loss component before the producer level is not usually shown on energy flow diagrams). Note: Some of the light energy absorbed through accessory pigments such as carotenoids widens the spectrum that can drive photosynthesis. However, much of accessory pigment activity is associated with photoprotection; they absorb and dissipate excess light energy that would otherwise damage chlorophyll.

6. (a) 78 835 kJ
   (b) $78\ 835 \div 1\ 700\ 000 \times 100 = 4.64\%$

7. (a) Decomposers and detritivores
   (b) Transport by wind or water to another ecosystem (e.g. blown or carried in water currents).

8. (a) Low oxygen or anaerobic, low temperature, low moisture.
   (b) Energy remains locked up in the detrital material and is not released.
   (c) Geological reservoir:

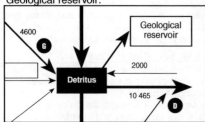

   (d) Oil (petroleum) and natural gas, formed from the remains of marine plankton. Coal and peat are both of plant origin; peat is partly decomposed, and coal is fossilised.

9. (a) $87\ 400 \rightarrow 14,000$: $14\ 000 \div 87\ 400 \times 100 = 16\%$
   (b) $14\ 000 \rightarrow 1600$: $1600 \div 14\ 000 \times 100 = 11.4\%$
   (c) $1600 \rightarrow 90$: $90 \div 1600 \times 100 = 5.6$
   (d) Producers to primary consumers

28. Agriculture and Productivity (page 35)
1. Farmers face challenges to maximise energy flow to the production of harvestable yield while minimising energy losses to pests and diseases, to ensure sustainability of the practice, and (in the case of animals) to meet certain standards of welfare and safety.

2. Pesticides can maximise the amount of energy captured (by reducing leaf loss) and converted to plant biomass.

3. (a) Antibiotics are used to combat diseases in a crowded environment thus resulting in greater production.
   (b) Prolonged use of low dose antibiotics encourages antibiotic resistance in the animals (or their gut microbes). This can make it harder to treat disease in livestock and contributes to the spread of antibiotic resistance.

4. Intensive crop farms are often monocultures which increase yield by maximising the amount of energy flow through a reduced number of trophic levels and decreased competition. Farmers also utilize pesticides and fertilisers to control pests and diseases and herbicides to control weeds (that would intercept light energy that could be used by the crop). In contrast, natural grasslands lose a greater amount of energy to pests and diseases (e.g. through loss of photosynthetic tissue). They must also compete for water and nutrients with other plants (weeds) which slows growth and production of biomass. Thus, their biomass is often much less than that of an intensive crop farms.

5. The efficiencies of intensive monocultures are achieved as a result of many high energy inputs (fertilisers, pesticides, energy costs of cultivation and irrigation). When these costs are accounted for, the natural system may be more productive.

6. (a) A farmer could create a monoculture to maximise the primary productivity of a particular crop, apply fertiliser, and control for pests and diseases.
   (b) Higher productivity in feed crops also allows greater secondary productivity (energy from feed crops is efficiently assimilated by livestock).

29. Nutrient Cycles (page 37)
1. (a) Bacteria are able to make conversions to and from elements and their ionic states. This gives plants and animals access to nutrients that they would otherwise not have (i.e. increases bioavailability).
   (b) Fungi decompose organic matter, returning nutrients to the soil where plants and bacteria can access them. They are also able to convert some nutrients into more readily accessible forms.
   (c) Plants are able make their own food and, when they die, add this to the soil in the form of nutrients that can be broken down and used by bacteria and fungi. They also provide browsing animals with nutrients when they are eaten.
   (d) Animals break down materials from plants, fungi and bacteria and return then to the soil with their wastes and when they die allowing the nutrients in them to re-enter the cycle.

2. The rates of decomposition are very high in the higher temperatures of tropical forests. As a result, decaying matter is processed very quickly and very little remains in the soil. Much of the carbon and other nutrients are also locked up in biomass.

3. A macronutrient is one that is required in large amounts. Micronutrients (also called trace elements) are needed in much smaller amounts.

30. The Nitrogen Cycle (page 38)
1. (a)-(e) any of:
   - Decomposition or decay of dead organisms, to ammonia by decomposer bacteria (ammonification).
   - Nitrification of ammonium ions to nitrite by nitrifying bacteria such as *Nitrosomonas* ($NH_4^+ \rightarrow NO_2^-$)
   - Nitrification of nitrite to nitrate by nitrifying bacteria such as *Nitrobacter* ($NO_2^- \rightarrow NO_3^-$)
   - Denitrification of nitrate to nitrogen gas by anaerobic denitrifying bacteria such as *Pseudomonas* ($NO_3^- \rightarrow N_{2(g)}$)
   - Fixation of atmospheric nitrogen to nitrate by nitrogen fixing bacteria such as *Azotobacter* and *Rhizobium* ($N_2 \rightarrow NO_3^-$)
   - Fixation of atmospheric nitrogen to ammonia by nitrogen fixing cyanobacteria ($N_2 \rightarrow NH_3$)
2. (a) Oxidation of atmospheric nitrogen by lightning.
   (b) Nitrogen fixation (by bacteria).
   (c) Production of nitrogen fertiliser by the Haber process.

3. Denitrification.

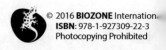
© 2016 **BIOZONE** International
ISBN: 978-1-927309-22-3
Photocopying Prohibited

4. The atmosphere.

5. Nitrate.

6. Any one of: Amino acids, proteins, chlorophyll.

7. Animals obtain their nitrogen by ingesting (eating) food (plants or other animals).

8. Leguminous material is high in nitrogen. Ploughing it in replenishes soil nitrogen and reduces the need for additional nitrogen fertiliser when growing non-leguminous crops subsequently.

9. (a)-(e) any five in any order:
   - Addition of nitrogen fertilisers to the land. This supplies inorganic nitrogen, as nitrate, for plant growth, but excess nitrogen, not absorbed by plants, may enter and pollute water sources.
   - Industrial physical-chemical fixation of nitrogen (through the Haber process) combines H and N to ammonia, which can be used to manufacture inorganic nitrogen fertilisers. This is an industrial process, requiring high temperatures and pressures and a large amount of energy. The effects of applied inorganic nitrogen are outlined above.
   - Genetic modification of plants so that they can fix nitrogen. The effect of this is to increase the range of crop plants capable of growing on nitrogen deficient soils. Potentially, this could make a beneficial contribution to soil fertility.
   - Large-scale, assisted composting produces nitrogen rich organic fertiliser which has the effect of improving soil fertility and structure. This has beneficial effects in reducing the amount of inorganic nitrogen fertiliser that must be applied for the desired plant growth.
   - Burning and harvesting removes nitrogen from the land and releases nitrogen oxides into the air.
   - Discharge of effluent (particular animal waste) into waterways enriches water bodies and leads to localised pollution and eutrophication.
   - Irrigation can accelerate loss of nitrate from the soil by increasing the rate at which nitrates are washed out of the soil into ground water.

31. Nitrogen Pollution (page 40)

1. (a) NO contributes to the formation of low level ozone which is a constituent of photochemical smog.
   (b) N_2O depletes ozone in the upper atmosphere.
   (c) NO_2 is a toxic inhalant. It also contributes to the formation of nitric acid in the atmosphere and therefore acid rain.
   (d) NO_3- is a water pollutant. It causes eutrophication, the accelerated growth of algae in waterways, and can cause severe health problems if drinking water contains significant amounts.

2. NO persists in the atmosphere both causing and being released by cyclic chemical reactions. NO reacts with oxygen to form toxic NO_2, which reacts with water to form HNO_3 (acid rain) and HNO_2. The HNO_2 decomposes, releasing NO to react again. NO will continue this cycling until it reacts with a chemical that removes it from the cycle.

3. Even after nitrogen fertilisers are not used, there is a large nitrogen load in soil and groundwater. Groundwater may take many years to move from its point of origin to its point of exit. Nitrate fertilisers that leach into groundwater now will move with this ground water and exit into waterways many years afterwards. In some cases, the lag may be up to fifty years.

4. (a) 1860: Reactive N deposition in the ocean = 156.5 million t. Release of unreactive nitrogen = 301 million t. 1995: reactive N deposition in the ocean = 202 million t. Release of unreactive nitrogen = 322 million t. This is an increase of 45.5 million t of reactive N deposition but an increase of only 21 million t of unreactive nitrogen released. Result: twice as much reactive nitrogen has been added to the ocean than has been released as unreactive nitrogen.
   (b) Algal blooms are becoming more common in the oceans as nitrate levels slowly rise. Many of these blooms are from algae that contain small amounts of toxins. These can be concentrated by filter feeders such as mussels and if eaten can cause poisoning.

5. (a) Nitrates are highly soluble in water and a lot is rapidly washed away or leached from the soil and not incorporated into plant tissues. Nitrates are also broken down by bacteria and returned to the air. Some nitrates will accumulate in the soil over time but not be accessible to plants. All these factors contribute to "lost" nitrogen.
   (b) Nitrogen losses could be minimised by fertiliser application at appropriate times and rates, and by sensible irrigation practices. Using slow release fertilisers in times of frequent rain also slows down the rate at which nitrates are lost into groundwater.

32. The Phosphorus Cycle (page 42)

1. Draw arrows from guano deposits and rock phosphate (labelled **mining/removal**) to: "Dissolved phosphates available to plants (PO_4^{3-})".

2. (a) – Mineralisation, i.e. the microbial conversion of organic P to forms available to plants (H_2PO_4 or HPO_4).
      – Immobilisation, i.e. microbial conversion of inorganic P to organic forms that are unavailable to plants. The microbial P will become available over time as the microbes die.
   (b) Any two of: DNA, ATP, phospholipids.
   (c) – Rock phosphate: Much phosphate is washed into the ocean where it builds up in phosphate-rich rocks made from marine sediments.
      – Bone deposits: Remains of dead marine vertebrates washed down rivers into lakes and into the sea.
      – Guano deposits: The droppings of birds (especially fish-eating birds) accumulated at nesting colonies. Cave dwelling bats also produce guano deposits.

3. Geological uplift and weathering (erosion).

33. Chapter Review (page 43)
No model answer. Summary is the student's own.

34. KEY TERMS AND IDEAS: Did You Get It? (page 44)

1. (a) water + carbon dioxide + light energy → glucose + oxygen
      $6H_2O + 6CO_2$ + light energy → $C_6H_{12}O_6 + 6O_2$
   (b) Photosynthesis occurs in the chloroplasts of green plants and algae (and the inner membranes of some photosynthetic bacteria).

2. (a) glucose + oxygen → carbon dioxide + water + ATP
      $C_6H_{12}O_6 + 6O_2 → 6CO_2 + 6H_2O$ + ATP
   (b) Respiration occurs in the mitochondria of all eukaryotic cells (and the inner membranes of some bacteria).

3. Absorption spectrum (G), aerobic respiration (H), ATP (P), action spectrum (I), Calvin cycle (D), cellular respiration (K), chlorophyll (J), food chain (O), gross primary productivity (N), Krebs cycle (B), light dependent phase (F), photosynthesis (A), stroma (E), thylakoid discs (C), trophic efficiency (L), trophic level (M).

4. (a) Primary production decreases with depth.
   (b) Photosynthesis occurs in the photic zone. Producers in the photic zone provide the basis of the marine food chain (most of the primary and higher order consumers), so most marine life is found here.

35. Detecting Changing States (page 47)

1. Any physical or chemical change in the environment, capable of provoking a response in an organism. Stimuli are detected by sense organs.

2. (a) and (b) any from the examples provided (or others). Answers provided as stimulus (receptor):
   (a) External stimuli and their receptors: Light (photoreceptor cells); gravity (hair cells/vestibule of the inner ear); sound/ vibration (hair cells in the cochlea of the inner ear); airborne chemicals (olfactory receptors in the nose); dissolved chemicals (taste buds); external pressure (dermal pressure receptors, e.g. Pacinian corpuscle, Meissner's corpuscles); pain (nerve endings in the dermis); temperature (nerve endings in the skin).
   (b) Internal stimuli and their receptors: Blood pH/carbon

dioxide level (chemoreceptors in blood vessels); blood pressure (baroreceptors); stretch (proprioreceptors, e.g. muscle spindle).

36. Plant Responses (page 48)

1. Light (including the light/dark cycle), gravity, temperature, touch, chemicals.

2. Appropriate responses to environmental stimuli enhance survival in different environments. They enable the plant to synchronise its daily cycles and seasonally important events, such as flowering and germination, with environmental cues.

3. (a) Closing of stomata
   (b) Leaf fall, dormancy
   (c) Leaf closure, leaf toxins
   (d) Closing of flowers

37. Tropisms and Growth Responses (page 49)

1. (a) Positive chemotropism
   (b) Negative gravitropism
   (c) Positive hydrotropism
   (d) Positive phototropism
   (e) Positive gravitropism
   (f) Positive thigmomorphogenesis (*alt.* thigmotropism)

2. (a) Enables roots to grow towards the soil (a suitable growing environment).
   (b) Enables coleoptiles to turn up and grow towards the sunlight (necessary for food manufacture).
   (c) Enables the plant to clamber upwards and grow toward the light instead of possibly becoming smothered by more upright plants.
   (d) Enables pollen tube to locate the micropyle of the embryo sac, and sperm nuclei to fertilise the egg.

3. Tropisms show adaptive value because they help position a plant in the most favourable conditions. For example, positive phototropism orientates seedlings towards the light and helps them obtain enough light for photosynthesis.

38. Auxins and Shoot Growth (page 50)

1. Positive phototropism

2. In an experiment in which a cut stem with an auxin infused agar block is uppermost and an agar block without auxin is at the base, auxin moves down the stem. However, if the system is inverted, no auxin is found in the stem, indicating that the auxin in the agar was not transported or diffused through the stem - it only travels one way.

3. Auxin causes cell elongation.

39. Investigating Phototropism (page 51)

1. (a) Auxin.
   (b) Positive phototropism.
   (c) **Point A**: Cells stay short.
       Point B: Cells elongate.
   (d) Side B
   (e)

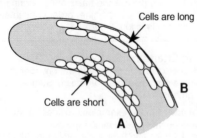

2. The hormone is produced in the shoot tip. The light initiates the response.

3. **Plant A**: The plant will exhibit phototropism and bend towards the sun.
   Plant B: The plant will exhibit no phototropic behaviour and will not bend.

40. Investigating Gravitropism (page 52)

1. (a) In shoots, more auxin accumulates on the lower side of the shoot. In response to higher auxin levels here, the cells on the lower side of the stem elongate and the shoot tip turns up.
   (b) In roots, the accumulation of auxin on the lower side inhibits elongation (since this is the response of roots to high auxin). The cells on the upper side therefore elongate more than those on the lower side and the root tip turns down.

2. (a) Approx. $10{-}3$ mg L^{-1} (b) Stem growth is promoted.

3. (a) A negative geotropic response ensures shoots turn up towards the light (important when light may be absent as when buried deeply in soil).
   (b) Positive geotropism ensures roots turn down into the soil so that they can begin obtaining the water and minerals required for continued growth.

41. Investigation of Gravitropism in Seeds (page 53)

1. (a) Down (towards the ground).
   (b) The root began to curve towards the ground.
   (c) The direction of the pull of gravity relative to the root had changed and so the root curved to compensate for the change of direction and continue to grow down.
   (d)

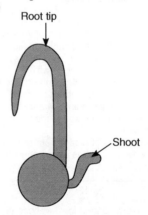

2. (a) Up (away from the ground)
   (b) The shoot continued to grow away from the ground.
   (c) The shoot reorientated itself to grow away from the ground, as this is the direction in which it is most likely to find light.

42. Taxes and Kineses (page 54)

1. A kinesis describes movement of a cell or organism in which the rate of movement depends upon the intensity (rather than direction) of the stimulus. An example is the increased activity of body lice when temperature increases over 30°C. A taxis is a directional movement in response to an external stimulus. An example is the negative phototaxis of maggots.

2. Simple orientation behaviours operate to position the animal in an environment that is favourable to its survival.

3. (a) Gravi –   Gravity
   (b) Hydro –   Water/moisture/humidity
   (c) Thigmo –   Touch
   (d) Photo –   Light
   (e) Chemo –   Chemical
   (f) Thermo –   Heat

4. (a) Snails: negative gravitaxis.
   (b) Moth: positive chemotaxis.
   (c) Louse: kinesis in response to temperature.
   (d) Lobster: positive thigmotaxis.
   (e) Mosquitoes: positive thermotaxis.
   (f) Maggots: negative phototaxis.

5. (a) i.   0-10%
       ii.   0-10%
       iii. 90-100%
   (b) High numbers of turns and fast movements return the woodlice to an area of high humidity.
   (c) 80-100%

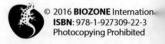
© 2016 **BIOZONE** International
ISBN: 978-1-927309-22-3
Photocopying Prohibited

6. (a) Nematodes move towards the NH_4Cl in plate A but make random movements in plate B.
   (b) Positive chemotaxis
   (c) Greater ability to find food and avoid toxins.

---

44. Choice Chamber Investigation (page 57)

1. H_0 for test 1:
   Woodlice have no preference of light or dark environments.

2. (a)

Category	Number of woodlice in chamber				
	O	E	O-E	$(O-E)^2$	$(O-E)^2/E$
Dark	33	20	7	49	2.45
Light	7	20	-13	169	8.45

$\chi^2 = 2.45 + 8.45$
$\chi^2 = 10.9$
   (b) 2-1 = 1

3. The test is significant at $P=0.05$. We can reject H_0 in favour of the alternative hypothesis that there is a significant difference in preference for the two environments.

4. H_0 for test 2:
   Woodlice have no preference of warm or cool environments.

5. (a)

Category	Number of woodlice in chamber				
	O	E	O-E	$(O-E)^2$	$(O-E)^2/E$
Warm	8	20	-12	144	7.2
Cool	32	20	8	64	3.2

$\chi^2 = 7.2 + 3.2$
$\chi^2 = 10.4$
   (b) 2-1 = 1

6. The test is significant at $P=0.05$. We can reject H_0 in favour of the alternative hypothesis that there is a significant difference in preference for the two environments.

7. Woodlice prefer dark and cool environments.

---

45. Reflexes (page 58)

1. Higher reasoning is not a preferable feature of reflexes because it would slow down the response time. The adaptive value of reflexes is in allowing a very rapid response to a stimulus.

2. A spinal reflex involves integration within the spinal cord, e.g. knee jerk (monosynaptic) or pain withdrawal (polysynaptic). A cranial reflex involves integration within the brain stem (e.g. pupil reflex).

3. (a) The knee jerk reflex helps to maintain posture and balance when walking. Being a reflex it requires no conscious though and means the brain is not devoting energy to correcting posture while walking.
   (b) The blink reflex protects the eye against damage by foreign bodies.
   (c) The grasp reflex in infants ensures the infant remains attached to its mother (e.g. grasping fur in primates or a hand in humans).
   (d) The pupillary light reflex regulates light entering the retina and ensures bright light does not damage the retina.

---

46. The Basis of Sensory Perception (page 59)

1. Sensory receptors convert stimulus energy (e.g. electromagnetic radiation) into electrochemical energy (a change in membrane potential).

2. All receptors receive and respond to stimuli by producing receptor potentials.

3. The stimulus energy opens an ion channel in the membrane leading to ion flux and a localised change in membrane potential, e.g. an influx of Na^+ and a depolarisation. This localised change in membrane potential is called a receptor

potential and it may lead directly or indirectly to an action potential.

---

47. Encoding Information (page 60)

1. (a) Stimulus strength is encoded by the frequency of action potentials.
   (b) Frequency modulation is the only way to convey information about the stimulus strength to the brain because action potentials are 'all or none' (information can not be communicated by variations in amplitude).

2. In the Pacinian corpuscle, stronger pressure produces larger receptor potentials, threshold is reached more rapidly, and action potential frequency is higher.

3. Sensory adaptation allows the nervous system to cease responding to constant stimuli that do not change in intensity. Constant, background sensory information can be ignored.

---

48. The Structure of the Eye (page 61)

1. (a) **Cornea**: Responsible for most of the refraction (bending) of the incoming light.
   (b) **Ciliary body**: Secretes the aqueous humour which helps to maintain the shape of the eye and assists in refraction.
   (c) **Iris**: Regulates the amount of light entering the eye for vision in bright and dim light.

2. (a) The incoming light is refracted (primarily by the cornea) and the amount entering the eye is regulated by constriction of the pupil. The degree of refraction is adjusted through accommodation (changes to the shape of the lens) so that a sharp image is formed on the retina.
   (b) Accommodation is achieved by the action of the ciliary muscles pulling on the elastic lens and changing its shape. Note: When the ciliary muscle contracts there is decreased pressure on the suspensory ligament and the lens becomes more convex (to focus on near objects). When the ciliary muscle relaxes there is increased tension on the suspensory ligament and the lens is pulled into a thinner shape (to focus on distant objects).

---

49. The Physiology of Vision (page 62)

1. (a) The retina is the region of the eye responsible for receiving and responding to light. It contains the pigment-containing photoreceptor cells (the rods and cones) which absorb the light and produce an electrical response. This response is converted by other cells in the retina into action potentials in the optic nerve.
   (b) The optic nerve is formed from the axons of the retinal ganglion cells, and carries the action potentials from the retina through the optic chiasma to the visual cortex in the cerebrum.
   (c) The central fovea is the region of the retina with the highest cone density where acuity is greatest (sharpest vision).

2.

Feature	Rod cells	Cone cells
Visual pigment(s):	Rhodopsin (no color vision)	Iodopsin (three types)
Visual acuity:	Low	High
Overall function:	Vision in dim light, high sensitivity	Color vision, vision in bright light

3. (a)-(c) in any order
   (a) **Photoreceptor cells** (rods and cones) respond to light by producing graded receptor potentials.
   (b) **Bipolar neurons** form synapses with the rods and cones and transmit the changes in membrane potential to the ganglion cells. Each cone synapses with one bipolar (=high acuity) whereas many rods synapse with one bipolar cell (=high sensitivity).
   (c) **Ganglion cells** synapse with the bipolar cells and respond with depolarisation and generation of action potentials. Their axons form the optic nerve.

4. (a) **Horizontal cells** are interconnecting neurons that

© 2016 **BIOZONE** International
ISBN: 978-1-927309-22-3

help to integrate and regulate the input from multiple photoreceptor cells. They give information about contrast.

(b) **Amacrine cells** form synapses with bipolar cells and work laterally to affect output from the bipolar cells, enhancing information about light level.

5. Several rod cells synapse with each bipolar cell. This gives poor acuity but high sensitivity. Each cone cell synapses with only one bipolar cell and this gives high acuity but poor sensitivity.

6. (a) A photochemical pigment is a molecule (e.g. contained in the membranes of the photoreceptor cells) that undergoes a structural change when exposed to light (and is therefore light-sensitive).
(b) Rhodopsin in rods and iodopsin in cones.

7. Light falling on the retina causes structural changes in the photopigments of the rods and cones. These changes lead to the development of graded electrical signals (hyperpolarisations) which spread from the rods and cones, via the bipolar neurons, to the ganglion cells. The ganglion cells respond by depolarisation and transmit action potentials to the brain.

50. Sensitivity (page 64)
Exemplar results for two point threshold test (in mm). Results will vary between individuals and trials, but the general trend of the data should be:

Forearm: 26 mm
Back of hand: 9 mm
Palm of hand: 7 mm
Fingertip: 1-2 mm
Lips: < 1 mm

1. Lips and/or fingertips.

2. Forearm.

3. The lips and/or fingertips need to be sensitive to carry out their functions (locating and tasting food, communication etc,). They therefore need to have a great number of receptors. The forearm requires many fewer as it is not involved in intricate tasks.

51. The Intrinsic Regulation of Heartbeat (page 65)
1. (a) Sinoatrial node: Initiates cardiac cycle through the spontaneous generation of action potentials.
(b) Atrioventricular node: Delays the impulse.
(c) Bundle of His: Distributes action potentials over the ventricles (resulting in ventricular contraction).
(d) Intercalated discs: Specialised junctions allowing electrical impulses to spread rapidly through the heart muscle (electrical coupling).

2. Delaying the impulse at the AVN allows time for atrial contraction to finish before the ventricles contract.

3. Being able to influence the basic rhythm of the heartbeat enables the body to respond appropriately to increased demand, e.g. increased need to supply oxygen and remove carbon dioxide when exercising or in a fight or flight situation.

4. (a) Epinephrine (adrenaline) increases the heart rate.
(b) Noradrenaline (norepinephrine).

52. Extrinsic Control of Heart Rate (page 66)
1. (a) Increased venous return: Heart rate increases.
(b) Release of adrenaline: Heart rate increases.
(c) Increase in blood CO_2: Heart rate increases.

2. These effects are brought about by the cardiovascular centre (sympathetic output via the cardiac nerve).

3. Physical exercise increases venous return.

4. (a) Cardiac nerve   (b) Vagus nerve

5. Increased stretch in the vena cava indicates increased venous return and cardiac output must increase to cope with the increase. Increased stretch in the aorta indicates increased cardiac output and heart rate decreases. The two responses keep cardiac output regulated according to the body's needs.

53. Investigating the Effect of Exercise (page 67)
1. (a) The graph depends on student's response to exercise.
(b) Students should see an increase in both heart rate and breathing rate during the exercise period.

2. (a) After 1 min of rest, there should be a fall in heart and breathing rate. After 5 min, both should have decreased further, and may have returned to pre-exercise levels.
(b) Once exercise is completed, the body's metabolic rate falls. The demand for energy and oxygen falls, and the heart rate and breathing rate will fall accordingly.

54. Neurones and Neurotransmitters (page 68)
1. A neurone processes and transmits information as electrical impulses from the point of stimulus to the point of reception.

2. To transmit signals from one neurone to the next (or from neurone to effector) across the synapse.

3. (a) Electricity changes the rate of depolarisation and changes the amount of neurotransmitter released.
(b) Neurotransmitter took time to move through the solution from one heart to the next, causing a delay.

4. (a) Myelination increases the speed of impulse conduction.
(b) Myelination prevents ion leakage across the neurone membrane. The current is carried in the cytoplasm so that the action potential at one node is sufficient to trigger an action potential at the next. Myelin also reduces energy expenditure since fewer ions overall need to be pumped to restore resting potential after an action potential passes.
(c) Faster conduction speeds enable more rapid responses to stimuli.
(d) 0-200 um diameter

55. Transmission of Nerve Impulses (page 70)
1. An action potential is a self-regenerating depolarisation (electrochemical signal) that allows excitable cells (such as muscle and nerve cells) to carry a signal over a (varying) distance.

2. (a) Neurones are able to transmit electrical impulses.
(b) Supporting cells are not able to transmit impulses.

3. (a) Depolarisation: Na^+ channels open and Na^+ ions flood into the cell.
(b) Repolarisation: Na^+ channels close, K^+ channels open and K^+ ions move out of the cell.

4. (1) When the neurone receives the threshold-level stimulus, the membrane briefly becomes more permeable to Na^+, which floods into the cell, resulting in a depolarisation.

(2) After the Na^+ influx, the Na^+ gates close and K^+ gates open, causing a brief hyperpolarisation before the resting potential is restored.

(3)   The hyperpolarisation means that for a short time (1-2 ms) the neurone cannot respond, so the impulse only travels in one direction (away from the stimulus).

5. Resting potential is restored by closure of the Na^+ channels and opening of the K^+ channels. K^+ moves out to restore the negative charge to the cell interior. All voltage activated gates close and the resting state is restored.

6. (a) Action potential travels by saltatory conduction, with depolarisation and action potential generation at the modes of Ranvier.
(b) Action potential spreads by local current (conduction is generally slower).

7. Because the refractory period makes the neurone unable to respond for a brief period after an action potential has passed the impulse can pass in only one direction along the nerve (away from the cell body).

56. Chemical Synapses (page 72)
1. (a) A synapse is a junction between the end of one axon and the dendrite or cell body of a receiving neurone. Note: A synapse can also occur between the end of one axon and a muscle cell (this is called the neuromuscular junction or

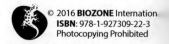

motor end plate).
(b) Cholinergic synapses are named for the neurotransmitter they release, i.e. acetylcholine.

2. Arrival of a nerve impulse at the end of the axon causes a calcium influx. This induces the vesicles to release their neurotransmitter into the cleft.

3. Delay is caused by the time it takes for the neurotransmitter to diffuse across the synaptic cleft.

4. The amount of neurotransmitter released influences the response of the receiving cell (response strength is proportional to amount of neurotransmitter released).

5. The response of the post-synaptic cell to the arrival of the neurotransmitter depends on the nature of the cell itself, on its location in the nervous system, and on the particular type of neurotransmitter involved.

6. (a) The neurotransmitter is deactivated by an enzyme (for acetylcholine this acetylcholinesterase).
(b) The neurotransmitter must be deactivated soon after its release to prevent continued stimulation of the post-synaptic cell (synaptic fatigue).
(c) The presynaptic neurone releases the neurotransmitter and the postsynaptic neurone has the receptors to bind and respond to that neurotransmitter. The information flow can therefore only be in one direction.

7. (a) They both involve release of the neurotransmitter acetylcholine in response to arrival of an action potential.
(b) Either of:
   - A neuromuscular junction is between a neurone and a muscle cell (fibre) not between two neurones.
   - A muscle fibre will respond maximally or not at all to ACh release, whereas the strength of the response in a postsynaptic cell to ACh is related to the amount of ACh released (synapses allow for synaptic integration including summation of inputs).

57. Integration at Synapses (page 74)

1. Integration refers to the interpretation and coordination (by the central nervous system) of inputs from many sources (inputs may be inhibitory or excitatory).

2. (a) and (b) any two of:
   - Chemical synapses transmit impulses in one direction to a precise location so information transfer is accurate.
   - They rely on a limited supply of neurotransmitter so they are subject to fatigue (inability to respond to repeated stimulation). This prevents the system from over stimulation.
   - Synapses also act as centres for the integration of inputs from many sources.

3. (a) Summation: The additive effect of presynaptic inputs (impulses) in the postsynaptic cell.
(b) Spatial summation refers to the summation of impulses from separate axon terminals arriving simultaneously at the postsynaptic cell. Temporal summation refers to the arrival of several impulses from a single axon terminal in rapid succession (the postsynaptic potentials are so close in time that they can sum to generate an action potential).

58. Drugs at Synapses (page 75)

1. (a) Drugs that decrease the usual effect of a neurotransmitter.
(b) Drugs that increase the usual effect of a neurotransmitter.

2. An antagonistic drug will decrease the inhibitory effect so the inhibition at the synapse will be lessened.

3. Atropine and curare are direct antagonists because they compete for the same binding sites as Ach on the postsynaptic membrane (direct) and block sodium influx so that impulses are not generated (hence antagonist).

4. An agonist that prevents neurotransmitter breakdown will increase the neurotransmitter's usual effect by prolonging its life at the synapse (stimulation of the postsynaptic cell continues for longer).

59. Antagonistic Muscles (page 76)

1. Muscles can only contract and relax, therefore they can only pull on a bone; they cannot push it. To produce movement, two muscles must act as **antagonistic** pairs to move a bone to and from different positions.

2. Muscles have an origin on one (less moveable) bone and an insertion on another (more moveable) bone. When the muscle contracts across the joint connecting the two bones, the insertion moves towards the origin, thereby moving the limb. To raise a limb, the flexor (prime mover in this case) contracts, pulling the limb bone up (extensor/antagonist relaxed). To lower the limb, the extensor contracts, pulling the limb down (flexor relaxed).

3. (a) Prime mover: the muscle primarily responsible for the movement.
(b) Antagonist: the muscle that opposes the prime mover, i.e. relaxes when prime mover contracts. Its action can be protective in preventing over-stretching of the prime mover during contraction.
(c) Synergist: assists the prime mover by fine-tuning the direction of limb movement.

4. Bones are rigid and movement occurs only at joints. The degree of movement allowed depends on the type of joint. The bones of the limbs, which need to be freely moveable are connected by synovial joints. Body parts where movement is not desirable, such as the joints of the skull, are largely rigid.

5. (a) Radius (radial tuberosity)
(b) Ulna
(c) Triceps (triceps brachii to give it its full name)
(d) Rotation of ulna and radius

6. Flexion of the knee joint would be a third class lever.

60. Skeletal Muscle Structure and Function (page 78)

1. (a) The banding pattern results from the overlap pattern of the thick and thin filaments (dark = thick and thin filaments overlapping, light = no overlap).
(b) I band: Becomes narrower as more filaments overlap and the area of non-overlap decreases.
H zone: Disappears as the overlap becomes maximal (no region of only thick filaments).
Sarcomere: Shortens progressively as the overlap becomes maximal.

2. (a) Relaxed
(b) The I band can clearly be seen. If the muscle was contracted the I band would not be present.

3. The all-or-none law of muscle contraction refers to the response of a single muscle fibre to stimulation which is to contract maximally (to threshold stimulation) or not at all.

4. (a) Changing the frequency of stimulation.
(b) Changing the number of fibres active at any one time.

61. The Sliding Filament Theory (page 80)

1. (a) Myosin: Has a moveable head that provides a power stroke when activated.
(b) Actin: Two protein molecules twisted in a double helix shape that form the thin filament of a myofibril.
(c) Calcium ions: Bind to the blocking molecules, causing them to move and expose the myosin binding site.
(d) Troponin-tropomyosin: Bind to actin molecule in a way that prevents myosin head from forming a cross bridge.
(e) ATP: Supplies energy for flexing of the myosin head (power stroke).

2. (a) ATP (hydrolysis) and Ca^{2+} (to expose the binding sites).
(b) Ca^{2+} is released from the sarcoplasmic reticulum. ATP is produced by cellular respiration in the mitochondria.

3. Muscle fibres use a lot of ATP during contraction. Having many mitochondria ensures adequate ATP supply.

62. Energy for Muscle Contraction (page 81)
1. Energy systems:

ATP-CP
ATP supplied by: Breakdown of CP
Duration of ATP supply: Short (3-15 s)

Glycolytic
ATP supplied by: Anaerobic breakdown of glycogen
Duration of ATP supply: A few minutes at most.

Oxidative
ATP supplied by: Complete aerobic (oxidative) breakdown of glycogen to CO_2 and water.
Duration of ATP supply: Prolonged but dependent on ability to supply oxygen to the muscles (fitness).

2. The glycolytic system can not supply enough ATP for prolonged activity. Continued anaerobic metabolism results in a build up of hydrogen ions which ultimately impedes muscle contraction.

3. (a) Oxidative metabolism relies on a continued supply of oxygen and on glucose, stored glycogen, or stored triglycerides for fuel.
   (b) As exercise intensity increases, readily available supplies of glucose decline and non carbohydrate sources must supply Krebs cycle. Free fatty acids from triglyceride breakdown are oxidised in the mitochondria and enter the Krebs cycle as 2C acetyl coenzyme A.

63. Muscle Fatigue (page 82)
1. Both fibre types generally produce the same force per contraction, but fast twitch fibres produce that force at a higher rate (power is the rate of doing work so this is equivalent to higher power output). Fast twitch fibres have a rapid rate of contraction and high power production but fatigue rapidly. Slow twitch fibres have a lower rate of contraction and a lower power output, but fatigue slowly.

2. (a) Oxygen is a limiting factor during intense exercise.
   (b) Hydrogen increases (acidosis) because protons are not being removed via the mitochondrial electron transport system (lack of oxygen as the terminal electron acceptor). Lactate accumulates faster than it can be oxidised and phosphate (from the breakdown of ATP and creatine phosphate) accumulates. These metabolic changes lead to a fall in ATP production and impaired calcium release from the sarcoplasmic reticulum, both of which contribute to inability of the muscle to do work (i.e. fatigue).

3. The maximum tension a muscle can produce under low pH conditions is half that can be produced at normal pH.

64. Investigating Muscle Fatigue (page 83)
1 (a) Graph based on student results. Generally the line should trend down over time.
   (b) Student results. Generally the number of times the peg is opened will decline over time.
   (c) Student answer as per results.
   (d) In most cases the number of peg openings will be lower.

65. Homeostasis (page 84)
1. Homeostasis is the relatively constant internal state of an organism, even when the external environment is changing.

2. (a) Detects a change in the environment and sends a message (electrical impulse) to the control centre.
   (b) Receives messages sent from the receptor, processes the sensory input and coordinates an appropriate response by sending a message to an effector.
   (c) Responds to the message from the control centre and brings about the appropriate response, e.g. muscle contraction or secretion from a gland.

66. Negative Feedback (page 85)
1. Negative feedback mechanisms are self-correcting (the response counteracts changes away from a set point) so that fluctuations are reduced. This stabilises physiological systems against excessive change and maintains a steady state.

2. A: Eating or food entering the stomach.
   B: Emptying of stomach contents.

67. Positive Feedback (page 86)
1. (a) Positive feedback has a role in amplifying a physiological process to bring about a particular response. Examples include (1) elevation in body temperature (fever) to accelerate protective immune responses, (2) positive feedback between oestrogen and LH to leading to an LH surge and ovulation, (3) positive feedback between oxytocin and uterine contractions: oxytocin causes uterine contraction and stretching of the cervix, which causes more release of oxytocin and so on until the delivery of the infant, (4) positive feedback in fruit ripening where ethylene accelerates ripening of nearby fruit.
   (b) Positive feedback is inherently unstable because it causes an escalation in the physiological response, pushing it outside the tolerable physiological range. Compare this with negative feedback, which is self correcting and causes the system to return to the steady state.
   (c) Positive feedback loops are normally ended by a resolution of situation causing the initial stimulation. For example, the positive feedback loop between oestrogen and LH leading to ovulation is initiated by high oestrogen levels and ended when these fall quickly after ovulation, prompting a resumption of negative feedback mechanisms. In childbirth, once the infant is delivered, the stretching of the cervix ceases and so too does the stimulation for more oxytocin release.
   (d) When positive feedback continues unchecked, it can lead to physiological collapse. One example includes unresolved fever. If an infection is not brought under control (e.g. by the body's immune system mechanisms or medical intervention), body temperature continues to rise, leading to seizures, neurological damage, and death.

68. Cell Signalling (page 87)
1. (a) **Endocrine signalling** involves a hormone being carried in the blood between the endocrine gland/organ where it is produced to target cells.
   (b) **Paracrine signalling** involves cell signalling molecules being released to act on target cells in the immediate vicinity, e.g. at synapses or between cells during development.
   (c) **Autocrine signalling** involves cells producing and reacting to their own signals (e.g. growth factors from T cells stimulate production of more T cells).

2. The three signalling types all have in common some kind of chemical messenger or signal molecule (ligand) and a receptor molecule (on the target cells, which may or may not be on the cell producing the signal).

69. Hormonal Regulatory Systems (page 88)
1. (a) Antagonistic hormones are two hormones that have contrasting (counteracting) effects on metabolism. Examples: insulin and glucagon, parathormone (increases blood calcium) and calcitonin (lowers blood calcium).
   (b) In general principle, the product of a series of (hormone controlled) reactions controls its own production by turning off the pathway when it reaches a certain level. If there is too little product, its production is switched on again.

2. Only the target cells have the appropriate receptors on the membrane to respond to the hormone. Other (non-target) cells will not be affected.

3. (a) Hormones must circulate in the blood to reach the target cells and a metabolic response must be initiated. This takes some time.
   (b) Hormones bring about a metabolic change and often start a sequence of cascading, interrelated events. Once started, these events take time to conclude. Nervous responses continue only for the time that the stimulation continues.

70. Control of Blood Glucose (page 89)

© 2016 **BIOZONE** International
ISBN: 978-1-927309-22-3
Photocopying Prohibited

1. (a) Stimulus: Rise in the levels of glucose in the blood above a set level (about 100 mg dL^{-1} or 5.6 mmol per L).
   (b) Stimulus: Fall in blood glucose levels below a set level (about 70 mg dL^{-1} or 3.9 mmol per L).
   (c) Glucagon brings about the production (and subsequent release) of glucose from the liver by the breakdown of glycogen and the synthesis of glucose from amino acids.
   (d) Insulin increases glucose uptake by cells and brings about production of glycogen and fat from glucose in the liver.

2. Fluctuations in blood glucose (BG) and blood insulin levels are closely aligned. Following a meal, BG rises sharply and there is a corresponding increase in blood insulin, which promotes cellular glucose uptake and a subsequent fall in BG. This pattern is repeated after each meal, with the evening meal followed by a gradual decline in BG and insulin over the sleep (fasting) period. Negative feedback mechanisms prevent excessive fluctuations in blood glucose (BG) throughout the 24 hour period.

3. Humoral (blood glucose level).

71. The Liver's Role in Carbohydrate Metabolism (page 90)

1. In any order:
   (a) Glycogenesis: the production of glycogen from glucose in the liver, stimulated by insulin.
   (b) Glycogenolysis: breakdown of glycogen to produce glucose, stimulated by adrenaline and glucagon.
   (c) Gluconeogenesis: the production of glucose from non-carbohydrate sources, stimulated by adrenaline and glucocorticoid hormones.

2. (a) 1: Glycogenesis (formation of glycogen from glucose).
   (b) 2: Glycogenolysis (glycogen breakdown).
   (c) 3: Gluconeogensis (formation of glucose from non-carbohydrate sources).

3. Interconversion of carbohydrates is essential to regulating blood glucose levels and maintaining a readily available supply of glucose as fuel without incurring the homeostatic problems of high circulating levels of glucose.

72. Insulin and Glucose Uptake (page 91)

1. Insulin is secreted by the β cells of the pancreatic islets.

2. (a) Too much insulin results in low blood glucose levels (hypoglycaemia).
   (b) Too little insulin results in elevated blood glucose levels (hyperglycaemia).

3. Two molecules of insulin must bind to the extracellular domain of the insulin receptor to activate it. Once the insulin is bound, phosphate groups are added to the receptor. This phosphorylation begins a signal cascade resulting in the activation of Glut4 secretory vesicles, which produce the Glut4 glucose transporters. The Glut4 glucose transporters insert into the membrane allowing the uptake of glucose.

4. In a signal cascade, the sequence (chain reaction) of phosphorylations results in the activation of an increasing number of other proteins. Thus for two signal molecules (insulin) many proteins are activated and many Glut4 secretory vesicles (each producing transporters) are made.

73. Adrenaline and Glucose Metabolism (page 92)

1. A water soluble first messenger cannot cross the plasma membrane as it is polar and repelled by the non-polar nature of the phospholipid tails in the plasma membrane. It must interact with a receptor on the membrane and activate a second messenger inside the cell.

2. Phosphorylation of the molecules causes a shape change that changes the molecule from an inactive form to an active form, i.e. phosphorylation activates inactive proteins (enzymes).

3. A first messenger binds to and activates a G-protein linked receptor. The G-protein is activated, in turn activating adenylate cyclase which catalyses the synthesis of cAMP (the second messenger). cAMP initiates the phosphorylation cascade that leads to a cellular response. The role of G proteins, therefore, is to act as a molecular switch, regulating

the activity of a signal transduction pathway.

74. Type 1 Diabetes Mellitus (page 93)

1. Hyperglycemia (high blood glucose) results from the inability of cells to take up glucose. Glucose is normally reabsorbed in the kidney tubule, but when blood glucose is too high, it exceeds the kidney's ability to reabsorb it from the filtrate and so glucose is excreted in the urine (glucosuria). Excessive thirst results from the high urine volumes associated with the excretion of excess glucose. Hunger, fatigue and weight loss result from the inability to utilise glucose. Ketosis results from the metabolism of fats (used because glucose is not entering the cell so cannot be used as an energy source).

2. Regular injections of insulin restore the levels of insulin in the blood, allowing the cells to take up glucose.

75. Type 2 Diabetes Mellitus (page 94)

1. Type 1 diabetes results from a non-production of insulin and must be treated with insulin injection. Type 2 results from the body's cells becoming insensitive to normal levels of insulin. It is treated first with dietary management and exercise. Insulin therapy is usually not involved except in severe cases.

2. An increase in obesity and inactivity both contribute to the risk of developing diabetes.

76. Measuring Glucose in a Urine Sample (page 95)

1. (a) Approximately 20 mg dL^{-1}
   (b) Approximately 10 mg dL^{-1}
   (c) Approximately 140 mg dL^{-1}

2. (a) Sample 3 gives the most cause for concern.
   (b) Urine ordinarily contains no glucose because it is reabsorbed by the proximal tubules of the kidney.

3. The 0 mg dL^{-1} serves as control for the experiment.

4. You would have to dilute your unknowns by a known amount and then recalculate based on your dilution factor. If your unknowns were too dilute, you would have to dilute to create a new set of standards, remeasure and produce a new calibration curve.

77. The Physiology of the Kidney (page 96)

1. (a) Renal corpuscle: Blood is filtered through the capillaries of the glomerulus.
   (b) Proximal convoluted tubule: ~90% of the glomerular filtrate is reabsorbed (including glucose and ions).
   (c) Loop of Henle: Transport of salt and osmosis produce a salt gradient through the kidney.
   (d) Distal convoluted tubule: Modification of filtrate by active reabsorption and secretion of ions.
   (e) Collecting duct: Osmotic withdrawal of water and concentration of filtrate (urine).

2. Alignment of nephrons (through the salt gradient in the kidney) allows urine to be concentrated as it flows towards the renal pelvis and ureter.

3. The high blood pressure is needed for ultrafiltration, i.e. to force small molecules such as water, glucose, amino acids, sodium chloride and urea through the capillaries of the glomerulus and the basement membrane and epithelium of Bowman's capsule.

4. (a) Glomerular filtration: Produces an initial filtrate of the blood that is similar in composition to blood and can be modified to produce the final urine.
   (b) Active secretion: Secretion allows the body to get rid of unwanted substances into the urine.
   Explanatory detail: Active secretion of chloride in the ascending limb (with sodium following passively) contributes to the maintenance of the salt gradient in the extracellular fluid (this gradient allows water to be reabsorbed in the collecting duct). Secretion of toxins and unwanted ions into the filtrate in the distal tubules allows the blood composition to be adjusted and poisons to be excreted. Energy is used to secrete these substances against their concentration gradients
   (c) Reabsorption: Essential process that allows the useful

© 2016 **BIOZONE** International
ISBN: 978-1-927309-22-3
Photocopying Prohibited

substances (required by the body) to be retained from the filtrate (particularly the initial filtrate, where 90% is reabsorbed). The body would waste energy if these substances were not retained.

5. (a) The salt gradient allows water to be withdrawn from the urine (allows the urine to be concentrated). **Explanatory detail**: Because the salt gradient increases through the medulla, the osmotic gradient is maintained and water can be continually withdrawn from the urine.
   (b) Salt gradient is produced by the active and passive movement of salt from the filtrate into the extracellular fluid in the medulla.

6. (a) 25 minutes after drinking the water, urine volume had nearly doubled. After 50 minutes urine volume had increased more than threefold from the starting (reference) volume of 100 cm^3. After this time, urine volume declined steadily, returning to the reference volume after 150 minutes.
   (b) There is a time lag between drinking the water and clearing the majority of the excess fluid from the body (the time taken for filtration of the blood and urine formation).

---

78. Osmoregulation (page 98)

1. ADH promotes the reabsorption of water from the kidney collecting ducts, producing a more concentrated urine.

2. Blood volume is kept within narrow limits by adjusting urine volume and concentration. When receptors in the hypothalamus detect low blood volume, urine volume decreases and urine concentration increases. When receptors in the hypothalamus detect high blood volume, urine volume increases and urine concentration decreases.

3. High fluid intake would decrease ADH production.

4. (a) Individuals with diabetes insipidus would produce large volumes of dilute (insipid) urine.
   (b) Individuals with diabetes insipidus would have to take a synthetic form of ADH to compensate for the lack of natural ADH.

---

79. Chapter Review (page 99)

No model answer. Summary is the student's own.

---

80. KEY TERMS AND IDEAS: Did You Get It? (page 101)

1. (a) Phototropism
   (b) Gravitropism
   (c) Nastic response (see activity 36)
   (d) Auxin

2. (a) Positive phototropism by the orchid shoot and negative phototropism by the roots.
   (b) Light is the stimulus involved.

3. (a)

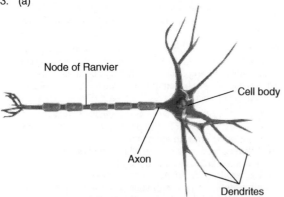

Node of Ranvier

Cell body

Axon

Dendrites

   (b) The neurone is myelinated.
   (c) The neurone shows areas of myelin alternating with nodes, which are not myelinated.
   (d) As action potentials by saltatory conduction.

4. (A) The membrane's resting potential.

(B) Membrane depolarisation (due to rapid Na$^+$ entry across the axon membrane).
(C) Repolarisation as the Na$^+$ channels close and slower K$^+$ channels begin to open.
(D) Hyperpolarisation (an overshoot caused by the delay in closing of K$^+$ channels).
(E) Return to resting potential after the stimulus has passed.

5. (a) Kidney
   (b) Nephron
   (c) Loop of Henle
   (d) Antidiuretic hormone (ADH)

6. Graph A
   Type: Positive feedback.

   Mode of action: Escalation (increase) of the physiological response away from the steady state condition. Once the outcome is achieved, the positive feedback mechanism ends.

   Biological examples: Fruit ripening, fever, labour, blood clotting.

   Graph B
   Type: Negative feedback.

   Mode of action: Has a stabilising effect. Maintains a steady state by counteracting variations from a set point.

   Biological examples: Most physiological processes are regulated by negative feedback, e.g. regulation of body temperature, blood glucose, calcium levels, and blood pressure and osmolarity.

7. Auxin (H), excretion (E), homeostasis (J), insulin (C), kidney (F), loop of Henle (B), muscle fibre (G), negative feedback (D), nephron (A), neuromuscular junction (K), positive feedback (M), reflex (N), sliding filament hypothesis (I), synapse (L).

---

81. Genotype and Phenotype (page 105)

1. (a) Sources include different alleles, sexual reproduction (recombination, independent assortment, random fertilisation), single nucleotide variations, and mutations.
   (b) Sources include nutrition, physical environment (e.g. temperature), drugs etc.

2. Because the environment also plays a role in the phenotype identical twins will always be slightly different. This includes fingerprints, height, and weight. Depending on the extent of the environment's influence, these differences may range from very slight or more obvious.

3. Behaviour (or components of it) can often be modified based on experience (learning). This means that the behavioural response can be dependent on the situation and the environment. For example, an animal can learn to use new food sources in different environments.

4. There are many examples of this. One example is the growth of the queen bee. A honey bee larva becomes a queen if fed royal jelly. If not it becomes a worker. Other examples include temperature dependent sex determination in reptiles, and colour-pointing in mammals.

---

82. Alleles (page 106)

1. (a) Heterozygous: Each of the homologous chromosomes contains a different allele for the gene (one dominant and one recessive).
   (b) Homozygous dominant: Each of the homologous chromosomes contains an identical dominant allele.
   (c) Homozygous recessive: Each of the homologous chromosomes contains an identical recessive allele.

2. (a) Aa   (b) AA   (c) aa

3. Each chromosome of a homologous pair comes from a different parent: one of maternal origin, one of paternal origin (they originated from the egg and the sperm that formed the zygote). They contain the same sequence of genes for the same traits, but the versions of the genes (alleles) on each chromosome may differ.

4. Alleles are different versions of the same gene that code for the same trait. Different alleles provide phenotypic variation for the expression of a gene. There are often two alleles

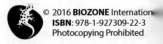 © 2016 **BIOZONE** International
ISBN: 978-1-927309-22-3
Photocopying Prohibited

for a gene, one dominant and one recessive. In this case, the dominant allele will be expressed in the phenotype. Sometimes alleles for a gene can be equally dominant, in which case, both alleles will be expressed in the phenotype. Where three or more alleles for a gene exist (multiple alleles), there is more phenotypic variation in the population (for that trait) than would be the case with just two alleles.

83. The Monohybrid Cross (page 107)

1. All the F_1 generation are heterozygous and thus display the dominant flower colour. When these are crossed together a 3:1 phenotypic ratio appears so that the homozygous recessive feature will be seen again.

2.

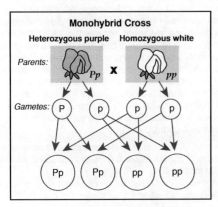

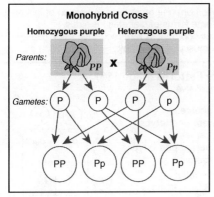

84. Practising Monohybrid Crosses (page 108)

1.

	Genotype	Phenotype
Cross 2	50% BB 50% Bb	100% black
Cross 3	25% BB 50% Bb	75% black
	25% bb	25% white
Cross 4	100% BB	100% black
Cross 5	50% Bb	50% black
	50% bb	50% white
Cross 6	100% bb	100% white

85. Codominance of Alleles (page 109)

1. Two or more alleles are dominant over any recessive alleles and both are fully expressed.

2. (a) Diagram labels:

	White bull	Roan cow
Parent genotype:	$C^W C^W$	$C^R C^W$
Gametes:	C^W, C^W	C^R, C^W
Calf genotypes:	$C^R C^W, C^W C^W$	$C^R C^W, C^W C^W$
Phenotypes:	roan, white	roan, white

(b) Phenotype ratio: 50% roan, 50% white (1:1)

(c) By breeding only from the roan calves. Offspring of roan parents should include white, roan, and red phenotypes. By selecting only the red offspring from this generation it would be possible to breed a pure herd of red cattle.

3. (a) Diagram labels:

	Unknown bull	Roan cow
Parent genotype:	$C^R C^R$	$C^R C^W$
Gametes:	C^R, C^R	C^R, C^W
Calf genotypes:	$C^R C^R, C^R C^W$	$C^R C^R, C^R C^W$
Phenotypes:	red, roan	red, roan

(b) Unknown bull: red bull

4. The phenotypic ratio would be 1 red: 2 roan: 1 white

86. Codominance in Multiple Allele Systems (page 110)

1. Blood group table:
   Blood group **B** BB, BO
   Blood group **AB** AB

2.

Cross 2	Group O	Group O
Gametes:	O, O	O, O
Children's genotypes:	OO, OO, OO, OO	
Blood groups:	O, O, O, O	
Cross 3	**Group AB**	**Group A**
Gametes:	A, B	A, O
Children's genotypes:	AA, AO, BA, BO	
Blood groups:	A, A, AB, B	
Cross 4	**Group A**	**Group B**
Gametes:	A, A	B, O
Children's genotypes:	AB, AO, AB, AO	
Blood groups:	AB, A, AB, A	
Cross 5	**Group A**	**Group O**
Gametes:	A, O	O, O
Children's genotypes:	AO, AO, OO, OO	
Blood groups:	A, A, O, O	
Cross 6	**Group B**	**Group O**
Gametes:	B, O	O, O
Children's genotypes:	BO, BO, OO, OO	
Blood groups:	B, B, O, O	

3. (a) Parental genotypes: AO, OO
       Gametes: A, O, O, O
       Children's genotypes: AO, AO, OO, OO

   (b) 1:2

   (c) 1:2

   (d) 0

4. (a) Man OO / Woman AA
       Children's genotypes: AO, AO, AO, AO
       Blood groups: A, A, A, A
       Man OO / Woman AO
       Children's genotypes: AO, AO, OO, OO
       Blood groups: A, A, O, O

   (b) The child's blood group is either A or O. Therefore, it is not possible for the man to be the child's father.

5. (a) Parental genotypes: AB, OO
       Gametes: A, B, O, O
       Children's genotypes: AO, AO, BO, BO
       Blood groups of children A or B

   (b) Possible parental genotypes: BB or BO, AA or AO
       Possible gamete combinations:
       B, B, A, A or B, O, A, A, or B, B, A, O, or B, O, A, O
       Blood groups of children: O, A, B, AB

87. Problems Involving Monohybrid Inheritance
(page 112)

1. 1/2 Ww and 1/2 ww.

   Ratio: 1 wire-haired : 1 smooth haired

   Working: Parental genotypes are Ww X Ww. The test cross of the F_1 (to the homozygous recessive by definition) is to a smooth haired dog (ww).

   1/4 of the F_1 will be wire-haired (WW). When crossed with ww the result will be all wire-haired dogs (Ww).

   1/2 the F_1 will be wire-haired (Ww). When crossed with ww, the result is 1/2 wire-haired and 1/2 smooth-haired.

   1/4 of the F_1 will be smooth-haired (ww). When crossed with ww, all offspring will be smooth-haired (ww). Across all progeny, half will be Ww and half will be ww.

2. Probability of black offspring: (2/3 x 1/4=) 1/6 or 0.16

   Working: The parents genotypes are Bb X Bb, and 1/3 of

© 2016 **BIOZONE** International
ISBN: 978-1-927309-22-3
Photocopying Prohibited

the white offspring (BB) crossed with Bb will result in no black lambs while 2/3 of the white offspring (Bb) crossed with Bb will result in 1/4 black lambs.

3. (a) They have an albino child (aa) as well as unaffected ones (AA or Aa), so the parents must both be Aa. Note: There is a 25% chance that any child of theirs will be albino.
   (b) The family are all aa.
   (c) The albino father must be aa. The mother must be Aa. The three unaffected children are Aa.
   Note: There is a 50% chance that any child of theirs will be albino. The observed 3:1 ratio is not surprising, given the small number of offspring.

4. – **Couple #1** genotypes must be X^HX- and X^HY because neither is affected. Their son is affected XhY. If the mother is X^HX^H they could not have an affected son. If she is X^HX^h, there is a 50% chance that her son will be XhY.
   – **Couple #2** genotypes must be X^HX- and X^hY and their son is X^HY. The father did not pass an X chromosome to his son, so his genotype is irrelevant. If the mother is X^HX^H, all of her sons will be X^HY, but if she is a carrier X^HX^h, there is a 50% chance that her son will be X^hY.
   – Either the hospital or the parents could be correct. The answer depends on the genotype of the mothers.

5. There is a possibility that the male is the father of the child as blood group O can result from crossing AO and BO genotypes. However there are also many other possible outcomes. Without more precise testing or knowing the actual genotypes of the male and female it is impossible to conclusively say the male is the father of the child.

88. Sex Linked Genes (page 113)

1. Parent genotype: X_oX_o X_oY
   Gametes: X_o, X_o X_o, Y
   Kitten genotypes: $X_oX_o, X_oY, X_oX_o, X_oY$

Genotypes	**Phenotypes**
Male kittens: X_oY	Black
Female kittens: X_oX_o	Tortoiseshell

2. Parent genotype: X_oX_o X_oY
   Gametes: X_o, X_o X_o, Y
   Kitten genotypes: X_oX_o, X_oX_o X_oY, X_oY
   Phenotypes: Orange female   Black male,
                        Tortoiseshell female   Orange male
   (a) Father's genotype: X_oY
   (b) Father's phenotype: Orange

3. Parent genotype: X_oX_o X_oY
   Gametes: X_o, X_o X_o, Y
   Kitten genotypes: X_oX_o, X_oY X_oX_o, X_oY
   Phenotypes: Tortoise female,   Tortoise female,
                        Black male   Black male
   (a) Father's genotype: X_oY
   (b) Father's phenotype: Orange
   (c) Yes, the same male cat could have fathered both litters.

4. Parent: Normal wife   Affected husband
   Parent genotype: XX X_RY
   Gametes: X, X X_R, Y
   Children's genotypes: X_RX, XY X_RX, XY
   Phenotypes: Affected girl,   Affected girl,
                        Normal boy   Normal boy
   (a) Probability of having affected children = 50% or 0.5
   (b) Probability of having an affected girl = 50% or 0.5
   However, all girls born will be affected = 100%
   (c) Probability of having an affected boy = 0% or none

5. Parent: Affected wife   Normal husband
   Parent genotype: X_RX XY
   Gametes: X_R, X X, Y
   Children's genotype: X_RX, X_RY XX, XY
   Phenotypes: Affected girl,   Normal girl,
                        Affected boy   Normal boy
   Note: Because the wife had a normal father, she must be heterozygous since her father was able to donate only an X-chromosome with the normal condition.
   (a) Probability of having affected children = 50% or 0.5
   (b) Probability of having an affected girl = 25% or 0.25
   However, half of all girls born may be affected.

(c) Probability of having an affected boy = 25% or 0.25
   However, half of all boys born may be affected.

Background for question 6: Sex linkage refers to the location of genes on one of the sex chromosomes (usually the X, but a few are carried on the Y). Such genes produce an inheritance pattern that is different from that shown by autosomes:

– Reciprocal crosses produce different results (unlike autosomal genes that produce the same results).
– Males carry only one allele of each gene.
– Dominance operates in females only.
– A 'cross-cross' inheritance pattern is produced: father to daughter to grandson, etc.

6. **Sex linkage** (in humans, this usually means X-linkage) is involved in a number of genetic disorders. X-linked disorders are commonly seen only in males, because they have only one locus for the gene and must express the trait. If the sex linked trait is due to a recessive allele, females will express the phenotype only when homozygous recessive. It is possible for females to inherit a double dose of the recessive allele (e.g. a colour blind daughter can be born to a colour blind father and mother who is a carrier), but this is much less likely than in males because sex linked traits are relatively uncommon. Over a hundred X-linked genes are known, including those that control:

– Blood clotting: A recessive allele for this gene causes haemophilia. It affects about 0.01% of males but is almost unheard of in females.
– Normal colour vision: A recessive allele causes red-green colour blindness affecting 8% of males but only 0.7% of females.
– Antidiuretic hormone production: A version of this gene causes some forms of diabetes insipidus.
– Muscle development: A rare recessive allele causes Duchene muscular dystrophy.

89. Inheritance Patterns (page 115)

1. Autosomal recessive:

   (a) Punnett square:
       Male parent phenotype:
       Normal, carrier
       Female parent phenotype:
       Normal, carrier
   (b) Phenotype ratio:
       Normal 3   Albino 1

	P	p
P	PP	Pp
p	Pp	pp

2. Autosomal dominant:

   (a) Punnett square:
       Male parent phenotype:
       Woolly hair
       Female parent phenotype:
       Woolly hair
   (b) Phenotype ratio:
       Normal 1   Woolly 3

	W	w
W	WW	Ww
w	Ww	ww

3. Sex linked recessive:

   (a) Punnett square:
       Male parent phenotype:
       Normal
       Female parent phenotype:
       Normal, carrier
   (b) Phenotype ratio:
       Females:
       Normal 2   Haemophiliac 0
       Males:
       Normal 1   Haemophiliac 1

	X	X_h
X	XX	XX_h
Y	XY	X_hY

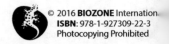
© 2016 **BIOZONE** International
ISBN: 978-1-927309-22-3
Photocopying Prohibited

4. Sex linked dominant:
   (a) Punnett square:
       Male parent phenotype:
       Affected (with rickets)
       Female parent phenotype:
       Affected (with rickets)

	X_R	X
X_R	$X_R X_R$	$X_R X$
Y	$X_R Y$	XY

   (b) Phenotype ratio:
       Females:
       Normal 0   Rickets 2
       Males:
       Normal 1   Rickets 1

90. Dihybrid Cross (page 116)

1.

	BL	Bl	bL	bl
BL	BBLL	BBLl	BbLL	BbLl
Bl	BBLl	BBll	BbLl	Bbll
bL	BbLL	BbLl	bbLL	bbLl
bl	BbLl	Bbll	bbLl	bbll

1 BBLL   2 BbLL   2 BBLl   4 BbLl
1 BBll   2 Bbll
1 bbLL   2 bbLl
1 bbll

91. Inheritance of Linked Genes (page 117)

1. Linkage refers to the situation where genes are located on the same chromosome. As a result, the genes tend to be inherited together as a unit.

2. Gene linkage reduces the amount of variation because the linked genes are inherited together and fewer genetic combinations of their alleles are possible.

3.

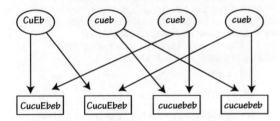

4. (a) CucuEbeb, Cucuebeb, cucuEbeb, cucuebeb
   (b) Offspring genotype: All CucuEbeb (heterozygotes)
       Offspring phenotype: All wild type (straight wing, grey body)

5. Female gametes = VgEb and vgEb, Male gametes = vgeb
   Offspring:
   Genotypes: VgvgEbeb   vgvgEbeb
   Phenotypes: straight wing, grey body. Vestigial wing, grey body

6. *Drosophila* produce a wide range of mutations, have a short reproductive cycle, produce large numbers of offspring and are easy to maintain in culture.

92. Recombination and Dihybrid Inheritance (page 119)

1. It produces new associations of alleles in offspring.

2.

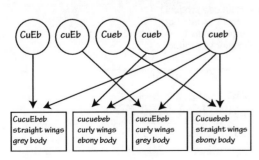

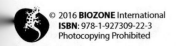

3. Parental linkage groups separate and new associations of alleles are formed in the offspring. The offspring show new combinations of characters that are unlike the parental types.

4. A greater than 50% recombination frequency indicates that there is independent assortment (the genes must be on separate chromosomes).

5. Female gametes: CuY , cuY
   Male gametes: Cuy, cuy
   Offspring genotypes and phenotypes:
   CuCuYy, CucuYy, CucuYy: Straight wings grey body
   cucuYy: Curly wings grey body

93. Detecting Linkage in Dihybrid Crosses (page 121)

1. Expected ratios as follows:
   Purple, long (P_L_)   215
   Purple, round (P_ll)   71
   Red, long (ppL_)   71
   Red, round (ppll)   24
   Total 381

2. (a) Expected ratios all 710 for each genotype as a 1:1:1:1 ratio is expected.
   (b) Parental (given), recombinant, recombinant, parental.
   (c) Morgan performed a test cross.

3. (a) Nail-patella syndrome is dominant. We can tell this because nearly all of the affected individuals had at least one parent with the disease.
   (b) The affected parent was blood group B. All of their offspring with the B blood group also had nail-patella syndrome. Therefore, nail-patella syndrome is linked to the B blood group allele.
   (c) Individual III-3 does not have nail-patella syndrome despite having the B blood type. It is likely that recombination has occurred, so this individual has not received the nail-patella gene.

95. Using the Chi-Squared Test in Genetics (page 123)

1. (a) H_0: "If both parents are heterozygous and there is independent assortment of alleles then we would expect to see a 9:3:3:1 ratio of phenotypes in the offspring".
   (b) H_A: "If both parents are heterozygous and the genes are linked (i.e. on the same chromosome), then we would expect the ratios of phenotypes in the offspring to deviate from the 9:3:3:1".

2. (a) Completed table:

Category	O	E	O − E	$(O − E)^2$	$\dfrac{(O − E)^2}{E}$
Purple stem, jagged leaf	12	16.3	−4.3	18.5	1.1
Purple stem, smooth leaf	9	5.4	3.6	13	2.4
Green stem, jagged leaf	8	5.4	2.6	6.8	1.3
Green stem, smooth leaf	0	1.8	−1.8	3.2	1.8
	Σ 29				Σ 6.6

Expected frequencies calculated as follows:
Purple stem, jagged leaf = 9/16 x 29 = 16.3
Purple stem, smooth leaf = 3/16 x 29 = 5.4
Green stem, jagged leaf = 3/16 x 29 = 5.4
Green stem, smooth leaf = 1/16 x 29 = 1.8
Note: Whole numbers could be used in preference to rounding to one decimal place.

   (b) $\chi^2 = 6.6$
   (c) Degrees of freedom = (4-1=) 3
   (d) The critical value of χ^2 at P = 0.05 and at d.f.= 3 is 7.82. The calculated χ^2 value is less than the critical value (6.6 < 7.82).
   (e) We cannot reject H_0: There was no significant difference between the observed results and the expected phenotype ratio of 9:3:3:1. We must conclude that the genes controlling stem colour and leaf shape in tomatoes are on separate chromosomes (unlinked).

3. (a) H_0 and H_A as for question 1.
    (b) Completed table:

Category	O	E	O – E	$(O-E)^2$	$\dfrac{(O-E)^2}{E}$
Round-yellow seed	441	450	–9	81	0.18
Round-green seed	159	150	9	81	0.54
Wrinkled-yellow seed	143	150	–7	49	0.33
Wrinkled-green seed	57	50	7	49	0.98
	Σ 800				Σ 2.03

Expected frequencies calculated as follows:
Round-yellow seed = 9/16 x 800 = 450
Round-green seed = 3/16 x 800 = 150
Wrinkled-yellow seed = 3/16 x 800 = 150
Wrinkled-green seed = 1/16 x 800 = 50
$\chi^2 = 2.03$

(c) Degrees of freedom = (4-1) 3.
    The critical value of χ^2 at P = 0.05 and at d.f.= 3 is 7.82.
    The calculated χ^2 is less than the critical value
    (2.03 < 7.82).
(d) We cannot reject H_0: There was no significant difference
    between the observed results and the expected
    phenotype ratio of 9:3:3:1. We must conclude that the
    genes controlling seed shape and colour are unlinked.

4. In both cases, we cannot reject H_0, but in the first case, the χ^2
    value is much higher. In tomatoes, the genes for stem colour
    and leaf shape are on separate chromosomes, but given the
    relatively large χ^2 value, repeating the experiment with more
    plants, or replicates, would serve as a check.

96. Problems Involving Dihybrid Inheritance (page 124)

1. (a)

	BL	Bl
Bl	BBLl	BBll
bl	BbLl	Bbll

Genotype ratio: 1BBLl: 1BBll: 1BbLl: 1Bbll
Phenotype ratio: 1 black short hair, 1: black long hair
(b)

	TL	Tl	tbL	tbl
tbL	TtbLL	TtbLl	tbtbLL	tbtbLl
tbl	TtbLl	Ttbll	tbtbLl	tbtbll

Genotype ratio:
1TtbLL : 2TtbLl : 1Ttbll : 1tbtbLL : 2tbtbLl : 1tbtbll
Phenotype ratio: 3: Tabby long hair, 1:Tabby short hair,
3: blotched tabby long hair, 1: blotched tabby short hair

2. (a) Self pollination of plants with orange striped flowers
        produces progeny in ratios close to 9:3:3:1 (the expected
        ratio of a cross between heterozygous offspring of true
        breeding parents). Thus you may hypothesise that O
        (orange petals) is dominant to o (yellow petals) and
        stripes (S) are dominant to no stripes (s).
    (b) The plants with the orange striped flowers were genotype
        OoSs. You know this because OoSs x OoSs will produce
        the progeny phenotypes in the observed 9:3:3:1 ratio.

3. (a) bbSS (brown/spotted) X BBss (solid/black)
        (which parent was male and which female is unknown.
        Parents must be homozygous since all the offspring are of
        one type: BbSs: black spotted).

(b) F_2 generation: BbSs X BbSs

	BS	Bs	bS	bs
BS	BBSS	BBSs	BbSS	BbSs
Bs	BBSs	BBss	BbSs	Bbss
bS	BbSS	BbSs	bbSS	bbSs
bs	BbSs	Bbss	bbSs	bbss

(c) Spotted/black   9/16
    Spotted/brown   3/16
    Solid/black   3/16
    Solid/brown   1/16
    Ratio: 9:3:3:1 (described as above)

4. (a) F_1: Genotype: all heterozygotes RrBb.
    (b) F_1: Phenotype: all rough black coats.
    (c) F_2 generation: RrBb X RrBb

	RB	Rb	rB	rb
RB	RRBB	RRBb	RrBB	RrBb
Rb	RRBb	RRbb	RrBb	Rrbb
rB	RrBB	RrBb	rrBB	rrBb
rb	RrBb	Rrbb	rrBb	rrbb

(d) Rough/black 9/16   Smooth/black 3/16
    Rough/white 3/16   Smooth/white 1/16
    Ratio: 9:3:3:1 (described as above)
(e)

	RB	Rb	rB	rb
RB	RRBB	RRBb	RrBB	RrBb

(f) F_2 Phenotype: all rough black coats.
(g) The parents' genotypes: RrBb X Rrbb

5. Note: Persian and Siamese parents are pedigrees
    (truebreeding) and homozygous for the genes involved.
    (a) Persian: UUss, Siamese: uuSS, Himalayan: uuss
    (b) F_1: Genotype: all heterozygotes UuSs.
    (c) F_1: Phenotype: all uniform colour, short haired.
    (d) F_2 generation: UuSs X UuSs

	US	Us	uS	us
US	UUSS	UUSs	UuSS	UuSs
Us	UUSs	UUss	UuSs	Uuss
uS	UuSS	UuSs	uuSS	uuSs
us	UuSs	Uuss	uuSs	uuss

(e) 1:15 or 1/16 uuss: Himalayan
(f) Yes (only one type of allele combination is possible)
(g) 3:13 or 3/16 (2 uuSs, 1 uuSS)

6. (a) Yes
    (b) Four phenotypes were produced. If there was no crossing
        over there would only be two phenotypes (parental types).
    (c) CucuEbeb, cucuebeb, Cucuebeb, cucuEbeb.

97. Gene Interactions (page 126)

1. (a) 9 walnut comb, 3 rose comb, 3 pea comb, 1 single comb.
    (b) Walnut comb
    (c) Single comb

2. (a) Colourless product
    (b) Colourless product
    (c) Purple product
    (d) Colourless product
    (e)

	AB	Ab
AB	AABB	AABb
Ab	AABb	AAbb
aB	AaBB	AaBb
ab	AaBb	Aabb

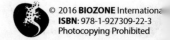

© 2016 **BIOZONE** International
ISBN: 978-1-927309-22-3
Photocopying Prohibited

3:1 purple to colourless

3. (a) GGYY, GgYY, GGYy, GgYy
   (b) ggYY, ggYy, GGyy, Ggyy, ggyy
   (c) Phenotypic ratio 9:7 Duplicate recessive epistasis

	GY	Gy	gY	gy
GY	GGYY	GGYy	GgYY	GgYy
Gy	GGYy	GGyy	GgYy	Ggyy
gY	GgYY	GgYy	ggYY	ggYy
gy	GgYy	Ggyy	ggYy	ggyy

4. Three colours are given, therefore the interaction must be **either recessive epistasis or dominant epistasis.**
   i. Any orange tailed fish crossed with any orange tailed fish produces only orange tailed fish. Therefore orange tailed fish could be (using the table) either BBhh or Bbhh, or bbHH, bbHh, or bbhh.
   ii. Red tailed fish must have dominant alleles because they produce all three colours when crossed with orange tailed fish (red is dominant to orange).
   iii. Assuming the interaction is recessive epistasis allows us to carry out crosses to test this. If we take BBhh to be orange or pink and bbHH to be pink or orange we get Bh x bH which gives BbHh - the heterozygote (red). Therefore the interaction cannot be recessive epistasis.
   To check we can assume the interaction is dominant epistasis. In this case if we take bbHH to be orange or pink and bbhh to be pink or orange we get bH x bh which gives bbHh - all the same colour (either pink or orange).
   To double check we can cross bbHh with bbhh:

	bH	bh
bh	bbHh	bbhh

Two different genotypes are produced. From the table we can see that the homozygous recessive produces one colour and the genotype containing a heterozygote is a different colour. (50%, 50%). The observation can then **lead us to conclude that the gene interaction is dominant epistasis.**

98. Epistasis (page 128)

1. No. of phenotypes: 3
2. Black:B_C_ (a dominant allele for each gene)
   Brown: A dominant allele for gene C only (e.g. Ccbb)
   Albino: No dominant allele for gene C (e.g. ccBB, ccbb)
3.

Eggs \ Sperm	BC	Bc	bC	bc
BC	**BBCC** Black	**BBCc** Black	**BbCC** Black	**BbCc** Black
Bc	**BBCc** Black	**BBcc** Albino	**BbCc** Black	**Bbcc** Albino
bC	**BbCC** Black	**BbCc** Black	**bbCC** Brown	**bbCc** Brown
bc	**BbCc** Black	**Bbcc** Albino	**bbCc** Brown	**bbcc** Albino

Ratio: Black: brown: albino
9: 3: 4

4. Homozygous albino (bbcc) x homozygous black (BBCC):
   Offspring genotype: 100% BbCc
   Offspring phenotype: 100% black
5. Homozygous brown (bbCC) x homozygous black (BBCC):
   Offspring genotype: 100% BbCC
   Offspring phenotype: 100% black
6. 4
7. Black: E_B_
   Brown: E_bb
   Yellow: eebb, eeB_
8. (a) All black, EeBb
   (b)

	EB	Eb	eB	eb
EB	EEBB black	EEBb black	EeBB black	EeBb black
Eb	EEBb black	EEbb brown	EeBb black	Eebb brown
eB	EeBB black	EeBb black	eeBB yellow (black nose)	eeBb yellow (black nose)
eb	EeBb black	Eebb brown	eeBb yellow (black nose)	eebb yellow (brown nose)

1:1:2:2:4:2:2:1:1
9 black, 3 brown, 3 yellow (black nose), 1 yellow (brown nose)

   (c) Collaboration (9:3:3:1)
9. (a) eeBb, eeBB
   (b) eebb
   (c) CCeeBb, CCeeBB

99. Polygenes (page 130)

1. Punnett square

Gametes	ABC	ABc	AbC	Abc	aBC	aBc	abC	abc
ABC	AABBCC	AABBCc	AABbCC	AABbCc	AaBBCC	AaBBCc	AaBbCC	AaBbCc
ABc	AABBCc	AABBcc	AABbCc	AABbcc	AaBBCc	AaBBcc	AaBbCc	AaBbcc
AbC	AABbCC	AABbCc	AAbbCC	AAbbCc	AaBbCC	AaBbCc	AabbCC	AabbCc
Abc	AABbCc	AABbcc	AAbbCc	AAbbcc	AaBbCc	AaBbcc	AabbCc	Aabbcc
aBC	AaBBCC	AaBBCc	AaBbCC	AaBbCc	aaBBCC	aaBBCc	aaBbCC	aaBbCc
aBc	AaBBCc	AaBBcc	AaBbCc	AaBbcc	aaBBCc	aaBBcc	aaBbCc	aaBbcc
abC	AaBbCC	AaBbCc	AabbCC	AabbCc	aaBbCC	aaBbCc	aabbCC	aabbCc
abc	AaBbCc	AaBbcc	AabbCc	Aabbcc	aaBbCc	aaBbcc	aabbCc	aabbcc

Darker skin | Same | Lighter skin

   (a) 20 (b) 27

2. Environmental influences will alter the colour of a person skin (such as tanning) to different extents.
3. Traits with continuous variation show a normal distribution curve when sampled and a graded variation in phenotype in the population. Such phenotypes are usually determined by a large number of genes and/or environmental influence. Examples include height, weight, hand span, foot size. In contrast, traits with discontinuous variation fall into one of a limited number of phenotypic variants and do not show a normal distribution curve when sampled. Differences in the phenotypes of individuals in a population are marked and do not grade into one other. Such phenotypes are usually controlled by a few alleles at a few genes, e.g. chin cleft.
4. Student's own plot. Shape of the distribution is dependent on the data collected. The plot should show a statistically normal distribution if sample is representative of the population and large enough.
   (a) Calculations based on the student's own data.
   (b) Continuous distribution, normal distribution, or bell shaped curve are all acceptable answers if the data conform to this pattern.
   (c) Polygenic inheritance: Several (two or more) genes are involved in determining the phenotypic trait. Environment may also have an influence, especially if traits such as

ISBN: 978-1-927309-22-3

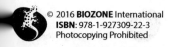

weight are chosen.

(d) A large enough sample size (30+) provides sufficient data to indicate the distribution. The larger the sample size, the more closely one would expect the data plot to approximate the normal curve (assuming the sample was drawn from a population with a normal distribution for that attribute).

100. Genes, inheritance, and Natural Selection

(page 132)

1. Mutation

2. Each individual has an effectively unique combination of alleles. Each individual is subjected to selection pressures that will affect its ability to survive and reproduce. In this way, the individual's allele combination is tested, with the better suited being able to produce more surviving offspring.

3. Mutation produces new alleles, which may contribute to increased or decreased fitness in the prevailing environment. Sexual reproduction, through independent assortment and random combination of gametes at fertilisation, shuffles alleles into new combinations. The new allele combinations, provide variation, the raw material on which natural selection acts. Selection acts on the phenotype (being the result of genotype and environment). Selection acts for or against particular allele combinations, enhancing or reducing the representation of these in the next generation. Selection therefore sorts the variability in a gene pool and establishes adaptive phenotypes.

101. Gene Pools and Evolution (page 133)

1. & 2. **Note**: Do not include the beetle about to enter Deme 1 (aa) but include the beetle about to leave Deme 1 (Aa). For the purpose of the exercise, assume that the individual with the mutation **A'A** in Deme 1 is a normal **AA**.

Deme 1: 22 beetles   Deme 2: 19 beetles

Deme 1		Number counted	%
Allele types	**A**	26	59.1
	a	18	40.9
Allele combinations	**AA**	8	36.4
	Aa	10	45.4
	aa	4	18.2

Deme 2		Number counted	%
Allele types	**A**	13	34.2
	a	25	65.8
Allele combinations	**AA**	1	5.3
	Aa	11	57.9
	aa	7	36.8

3. (a) **Population size**: Large population acts as a 'buffer' for random, directional changes in allele frequencies. A small population can exhibit changes in allele frequencies because of random loss of alleles (failure of an individual to contribute young to the next generation).

   (b) **Mate selection**: Random mating occurs in many animals and most plants. With 'mate selection', there is no random meeting of gametes, and certain combinations come together at a higher frequency than would occur by chance alone. This will alter the frequency of alleles in subsequent generations.

   (c) **Gene flow between populations**: Immigration (incoming) and emigration (outgoing) has the effect of adding or taking away alleles from a population that can change allele frequencies. In some cases, two-way movements may cancel, with no net effect.

   (d) **Mutations**: A source of new alleles. Most mutations are harmful, confer poor fitness, and will be lost from

the gene pool over a few generations. Some may be neutral, conferring no advantage over organisms with different alleles. Occasionally, mutations may confer improved fitness and will increase in frequency with each generation, at the expense of other alleles.

(e) **Natural selection**: Selection pressures will affect some allele types more than others, causing allele frequencies to change with each generation.

4. (a) Increase genetic variation: Gene flow (migration), large population size, mutation.

   (b) Decrease genetic variation: Natural selection, non-random mating (mate selection), genetic drift.

103. Modelling Natural Selection (page 137)

There is no set answer to this activity. Students should notice that the phenotypes that stand out from the background become reduced in number over time.

104. Changes in a Gene Pool (page 138)

1. This exercise (a)-(c) demonstrates how the allele frequencies change as different events take place:

Phase 1: Initial gene pool

This is the gene pool before any of the events take place:

	A	a	AA	Aa	aa
No.	27	23	7	13	5
%	54	46	28	52	20

Phase 2: Natural selection

The population is now reduced by 2 to 23. The removal of two homozygous recessive individuals has altered the allele combination frequencies (rounding errors occur).

	A	a	AA	Aa	aa
No.	27	19	7	13	3
%	58.7	41.3	30.4	56.5	13.0

Phase 3: Immigration / emigration

The addition of dominant alleles and the loss of recessive alleles makes further changes to the allele frequencies.

	A	a	AA	Aa	aa
No.	29	17	8	13	2
%	63	37	34.8	56.5	8.7

105. Hardy-Weinberg Calculations (page 139)

1. **Working**: $q = 0.1$, $p = 0.9$, $q^2 = 0.01$, $p^2 = 0.81$, $2pq = 0.18$
   Proportion of black offspring = $2pq + p^2$ x 100% = 99%;
   Proportion of grey offspring = q^2 x 100% = 1%

2. **Working**: $q = 0.3$, $p = 0.7$, $q^2 = 0.09$, $p^2 = 0.49$, $2pq = 0.42$
   (a) Frequency of tall (dominant) gene (allele): 70%
   (b) 42% heterozygous; 42% of 400 = 168

3. **Working**: $q = 0.6$, $p = 0.4$, $q^2 = 0.36$, $p^2 = 0.16$, $2pq = 0.48$
   (a) 40% dominant allele
   (b) 48% heterozygous; 48% of 1000 = 480.

4. **Working**: $q = 0.2$, $p = 0.8$, $q^2 = 0.04$, $p^2 = 0.64$, $2pq = 0.32$
   (a) 32% heterozygous (carriers)
   (b) 80% dominant allele

5. **Working**: $q = 0.5$, $p = 0.5$, $q^2 = 0.25$, $p^2 = 0.25$, $2pq = 0.5$
   Proportion of population that becomes white = 25%

6. **Working**: $q = 0.8$, $p = 0.2$, $q^2 = 0.64$, $p^2 = 0.04$, $2pq = 0.32$
   (a) 80%   (c) 36%   (e) 96%
   (b) 32%   (d) 4%

7. **Working**: $q = 0.1$, $p = 0.9$, $q^2 = 0.01$, $p^2 = 0.81$, $2pq = 0.18$
   Proportion of people expected to be albino (i.e. proportion that are homozygous recessive) = 1%

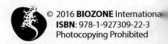

106. Analysis of a Squirrel Gene Pool (page 141)
1. Graph of population changes:

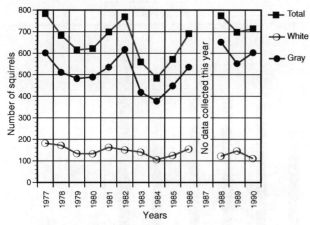

(a) 784 to 484 = 61% fluctuation
(b) Total population numbers exhibit an oscillation with a period of 5-6 years (2 cycles shown). Fluctuations occur in both grey and albino populations.

2. Graph of genotype changes:

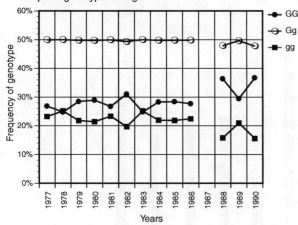

(a) GG genotype: Relatively constant frequency until the last 3-4 years, which show an increase. Possibly an increase over the total sampling period.
(b) Gg genotype: Uniform frequency.
(c) gg genotype: Relatively constant frequency until the last 3-4 years which exhibit a decline. Possibly a decrease over the total sampling period.

3. Graph of allele changes:

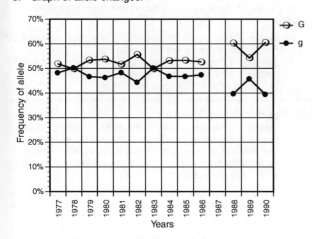

(a) Frequency of G: Increases in the last 3-4 years.
(b) Frequency of g: Decreases in the last 3-4 years.

4. (a) The *frequency of alleles* graph (to a lesser extent the *frequency of genotypes* graph)

(b) Changes in allele frequencies in a population provide the best indication of evolutionary change. These cannot be deduced simply from changes in numbers or genotypes.

5. There are at least two possible causes (any one of):
   – Genetic drift in a relatively small population, i.e. there are random changes in allele frequencies as a result of small population size.
   – Natural selection against albinos. Albinism represents a selective disadvantage in terms of survival and reproduction (albinos are more vulnerable to predators because of greater visibility and lower fitness).

107. Directional Selection in Moths (page 143)
1. The appearance of the wings and body (how speckled and how dark the pigmentation).

2. The selection pressure (the differential effect of selective predation on survival) changed from favouring the survival of light coloured forms in the unpolluted environments (prior to the Industrial Revolution) to favouring the dark morph (over the light morph) during the Industrial Revolution (when there was a lot of soot pollution). In more recent times, with air quality improving, the survival of the light coloured forms has once again improved.

3. As the frequency of the M allele increases so too does the frequency of the dark phenotype. Similarly as the frequency of the m allele decreases so too does the grey phenotype.

4. The frequency of the darker form fell from 95% to 50%.

108. Natural Selection in Pocket Mice (page 144)
1. (a) DD, Dd   (b) dd

2. See column graphs on the following page.

3. (a) The dark mice are found on the dark rocks.
   (b) Dark mice are found on the dark rocks as they blend in better, making it harder for predators to spot them. Dark mice on light rocks would be easily seen by predators. The same applies for light mice and dark rocks.
   (c) The mice at BLK and WHT do not conform to this generalisation. The mice at BLK are lighter than predicted from the lower rock reflectance and the mice at WHT are darker than predicted from the higher rock reflectance.
   (d) These mice might represent a recent migration to those areas from mostly dark or light population. There has not been enough time for the population to evolve to match their surroundings.

4. Selection is directional: in both environments, selective predation has caused a directional shift in the frequency of alleles controlling coat colour. The shift is similar to that seen in *Biston* moths.

5. Dark colour has evolved at least twice in rock pocket mice populations.

109. Disruptive Selection in Darwin's Finches (page 146)
1. (a) Large and small seeds became relatively more abundant.
   (b) The change in the relative abundance of seed sizes produced a negative selection pressure on finches with intermediate sized beaks. Those with smaller and larger beaks fared better during the drought because they could exploit the smaller and larger seed sizes.

2. Beak size determines fitness, which shows a bimodal distribution. Birds with small beaks (-1.0 single measure) or larger beaks (single measure 1.25) show higher fitness (leave more offspring) than birds with intermediate beak sizes.

3. (a) Mate selection is non-random.
   (b) The graph shows that birds tend to choose mates with a similar beak size. This relationship holds for wet and dry conditions.

110. Selection for Skin Colour in Humans (page 147)
1. (a) Folate is essential for healthy neural development.
   Explanatory note: A deficiency causes (usually fatal) neural tube defects (e.g. spina bifida).

Percentage reflectance of rock pocket mice coats

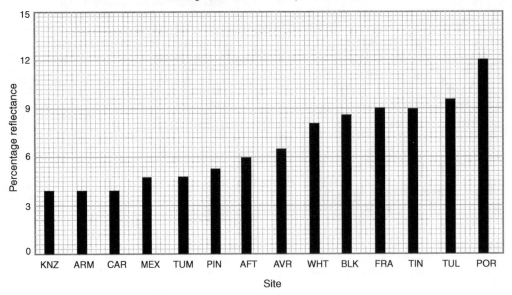

Percentage reflectance of rocks

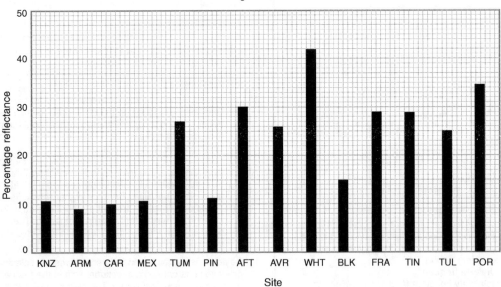

(b) Vitamin D is required for the absorption of dietary calcium and normal skeletal development. Explanatory note: A deficiency causes rickets in children or osteomalacia in adults. Osteomalacia in pregnancy can lead to pelvic fractures and inability to carry a pregnancy to term.

2. (a) Skin cancer normally develops after reproductive age and therefore protection against it provides no reproductive advantage and so no mechanism for selection.
   (b) The new hypothesis for the evolution of skin colour links the skin colour-UV correlation directly to evolutionary fitness (reproductive success). Skin needs to be dark enough to protect folate stores from destruction by UV and so prevent fatal neural defects in the offspring. However it also needs to be light enough to allow enough UV to penetrate the skin in order to manufacture vitamin D for calcium absorption. Without this, the female skeleton cannot successfully support a pregnancy. Because these pressures act on individuals both before and during reproductive age they provide a mechanism for selection. The balance of opposing selective pressures determines eventual skin colouration.

3. Women have a higher requirement for calcium during pregnancy and lactation. Calcium absorption is dependent on vitamin D, making selection pressure on females for lighter skins greater than for males.

4. The Inuit people have such abundant vitamin D in their diet that the selection pressure for lighter skin (for UV absorption and vitamin D synthesis) is reduced and their skin can be darker.

5. (a) Higher chances of getting rickets or (the adult equivalent) osteomalacia due to low UV absorption.
   (b) The simplest option to avoid these problems is for these people to take dietary supplements to increase the amount of vitamin D they obtain.

111. Genetic Drift (page 149)

1. (a) Genetic drift is the random changes in allele frequencies in populations which are unrelated to natural selection.
   (b) The effect of genetic drift is more pronounced in smaller populations than larger populations because of the relative size of the gene pools.

2. Genetic drift has a relatively greater effect in small populations and can result in the loss or fixation of alleles relatively rapidly. These changes can accumulate to a point that speciation occurs. Explanatory note: Unlike natural selection, genetic drift is a stochastic (random) process, and the changes due to genetic drift are not driven by environmental pressures. Changes in allele frequencies may therefore be beneficial, neutral, or detrimental to reproductive success.

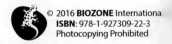
© 2016 **BIOZONE** International
ISBN: 978-1-927309-22-3
Photocopying Prohibited

112. The Founder Effect (page 150)

1.

Mainland	Nos	%		Nos	%
Allele A	48	54.5	Black	37	84
Allele a	40	45.5	Pale	7	16
Total	88	100			
Island	Nos	%		Nos	%
Allele A	12	75	Black	8	100
Allele a	4	25	Pale	0	0
Total	16	100			

2. The frequency of the dominant allele (A) is higher on the island population.

3. (a) Plants: Seeds are carried by wind, birds and water.
   (b) Animals: Reach islands largely by 'rafting' - situation where animals are carried offshore while clinging to vegetation; some animals survive better than others.
   (c) Non-marine birds: Blown off course and out to sea by a storm. Birds with strong stamina may survive.

4. Block A % for MDH-1 a in order 1-15: 61, 61, 64, 58, 61, 53, 68, 58, 56, 58, 56, 50, 50, 62, 25

   Block B % MDH-1 a in order 1-13: 19, 39, 25, 32, 30, 39, 30, 40, 42, 39, 46, 46, 53

5. Snails, although mobile, are restricted by their need for moisture. The tarmac roads and open areas on the blocks make movement between the two populations difficult as they are open and generally dry areas.

6. The frequencies have become different through a possible founder effect in which the original populations in each block had slightly different allele frequencies to begin with, or random events (i.e. genetic drift) have altered the proportion of alleles being passed to the next generation. Although adjacent, selection pressures in the blocks could also have been subtly different. It is likely that all these processes have contributed to the differences in allele frequencies.

7. Colony 15

113. Genetic Bottlenecks (page 152)

1. A sudden decrease in the size of a population can result in a corresponding reduction in genetic variation. This means the population has few 'genetic resources' to cope with the selection pressures imposed on it. In particular, it manifests itself as reduced reproductive success and a greater sensitivity to disease.

2. Poor genetic diversity means that if one individual is susceptible to a disease, then they are all likely to be vulnerable, a direct result of reduced genetic diversity.

3. With reduced genetic diversity, selection pressures on the population will have detrimental effects on fitness if one trait proves unfavourable. Since all cheetahs are virtually identical in their traits, individuals are similarly vulnerable and respond similarly to the same selection pressure.

114. Isolation and Species Formation (page 153)

1. Isolating mechanisms protect the gene pool from the diluting and potentially adverse effects of introduced genes. Species are well adapted to their niche; foreign genes will usually reduce fitness.

2. (a) Geographical isolation physically separates populations (and gene pools) but, if reintroduced, the two populations could potentially interbreed, i.e. reproductive isolation may not have occurred.
   (b) Geographical isolation enables populations to diverge in response to different selection pressures and (potentially) develop reproductive isolating mechanisms. Reproductive isolation won't generally occur in a populations in which there is gene flow (unless by special events such as polyploidy).

3. Geographical isolation physically separates populations (gene pools) so there is no gene flow between them. Ecological isolation arises as a result of different preferences in habitat or behaviour even though the populations occupy the same geographical area.

115. Reproductive Isolation (page 154)

1. (a) Postzygotic: hybrid breakdown
   (b) Prezygotic: structural
   (c) Prezygotic: temporal
   (d) Postzygotic: hybrid inviability

2. They are a secondary backup if the first isolating mechanism fails. The majority of species do not interbreed because of prezygotic mechanisms. Postzygotic mechanisms are generally rarer events.

116. Allopatric Speciation (page 156)

1. Animals may move into new environments to reduce competition for resources or because a new habitat becomes available (loss of geographical barrier or loss of another species freeing up an existing niche).

2. Plants move by dispersing their seeds.

3. Gene flow between the parent population and dispersing populations is regular.

4. Cooler periods (glacials) result in a drop in sea level as more water is stored as ice. As the temperature increases, the ice will begin to melt, and sea level will rise. The variation in sea level will depend on how much water is stored and released in response to the temperature change.

5. (a) Physical barriers that could isolate populations include the formation of mountain ranges, the formation of rivers or their change of course, the expansion or formation of desert, the advance of ice sheets, glacial retreat (isolating alpine adapted populations), and sea level rise. On a longer time scale, the formation of seas as a result of continental drift can isolate populations too.
   (b) Emigration (leaving one area and moving to another) will potentially reduce the genetic diversity of both gene pools, the migrants and the parent population. Depending on the extent of the migration, the effect will be the same as geographical isolation. The allele frequencies of the two isolates will diverge.

6. (a) The selection pressures on an isolated population may be quite different for that of the parent population. The immediate physical environment (e.g. temperature, wind exposure) as well as climatic region (e.g. temperate to tropical) may differ, as will biotic factors, such as competition, predation, and disease. In a different region, the food type and availability is also likely to different for the two populations. The shift in selection pressures may result in changes in allele frequencies as those best adapted to the new conditions survive to reproduce.
   (b) Some individuals in the isolated population will have allele combinations (and therefore a phenotype) that better suits the unique set of selection pressures at the new location. Over a period of time (generations) certain alleles for a gene will become more common in the gene pool, at the expense of other less suited alleles.

7. Reproductive isolation could develop in geographically isolated populations through the development of prezygotic and then postzygotic barriers to breeding. Prezygotic isolation would probably begin with ecological isolation, e.g. habitat preferences in the isolated population would diverge from the parent population in the new environment. Prezygotic isolating mechanisms that could develop subsequently to prevent successful mating include temporal isolation (e.g. seasonal shifts in the timing of breeding), incompatible behaviours (e.g. different mating rituals), and structural incompatibilities (e.g. incompatible mating apparatus). Gamete mortality (failure of egg and sperm to unite) can also prevent formation of the zygote in individuals that manage to copulate successfully. Once prezygotic isolation is established, post-zygotic mechanisms such as zygote mortality (in which the fertilised egg dies), reduced fertility in the hybrid, or hybrid breakdown (e.g. sterile F2) increase the isolation of the new species and prevent gene flow between it and the parent species.

8. Sympatric species are closely related species whose distribution overlaps. Allopatric species are species that remain geographically separated.

© 2016 **BIOZONE** International
ISBN: 978-1-927309-22-3
Photocopying Prohibited

117. Stages in Species Formation (page 158)

1. Some butterflies rested on top of boulders, others rested in the grass.

2. Selection pressure on BSBs is the need to maintain operating body temperature at the high altitude (fitness is higher when they can efficiently absorb heat from boulders). Selection pressure on the GSBs is probably predation as these lowland butterflies survive better where they can avoid detection.

118. Small Flies and Giant Buttercups (page 159)

1. When the original species of drosophilidae arrived on the Hawaiian islands it found many new unoccupied niches into which it expanded, resulting in a extensive adaptive radiation.

2. The fruit flies are of interest because there are so many closely related species within a small area and speciation has been relatively frequent. The flies also have a relatively simple genome, making genetic studies relatively easy.

3. In general the oldest species of flies are found on the oldest islands. As islands appeared out of the sea the flies spread to new environments and diversified, giving rise to newer species.

4. Buttercups living in alpine areas periodically have their habitats reduced and their range restricted during periods of climatic warming. This restricts gene flow and leads to speciation. Periods of cooling allow for the expansion of their range and movement to new environments as well as hybridisation to form new species. Repeated many times, these cycles lead to a large number of species.

119. Sympatric Speciation (page 160)

1. Sympatric speciation describes a speciation event that occurs without geographical separation. The two species are separated by some other means, such as niche differentiation or a spontaneous chromosomal change (polyploidy).

2. Polyploidy creates extra sets of chromosomes for an individual that make it impossible for it to reproduce with members of its parental population. Hybrids may form but they will be sterile.

3. Modern wheat, also seedless watermelons, kiwifruit

4. If two groups within a species population have slightly different habitats and food or foraging preferences (niche differentiation), then they will not come into contact for mating.

120. Components of an Ecosystem (page 161)

1. A community is a naturally occurring group of organisms living together as an ecological entity. The community is the biological part of the ecosystem. The ecosystem includes all of the organisms (the community) and their physical environment.

2. The biotic factors are the influences that result from the activities of living organisms in the community whereas the abiotic (physical) factors comprise the non-living part of the community, e.g. climate.

3. (a) Population   (c) The community
   (b) Ecosystem   (d) Physical factor

121. Types of Ecosystems (page 162)

1. (a) Yosemite National Park: ecosystem boundaries are the (artificial) boundaries of the park. It is part of the Sierra Nevada mountain range.
   (b) The clearing's border is the edge of the grass clearing where forest trees grow again.
   (c) The tree ecosystem borders are the tree and soil and air immediately around it.

122. Physical Factors and Gradients (page 163)

1. Environmental gradients:
   (a) Salinity: Increases from LWM to HWM
   (b) Temperature: Increases from LWM to HWM
   (c) Dissolved oxygen: Decreases from LWM to HWM
   (d) Exposure: Increases from LWM to HWM

2. (a) Mechanical force of wave action: Point B will receive the full force of waves moving inshore, Point A will receive only milder backwash, Point C will experience some surge but no direct wave impacts.
   Surface temperature: Points A and B will experience greater variations in rock temperature depending on whether the tide is in or out, day or night, water temperature, wind chill. Point C is more protected from some of these factors and will not experience the warming effect of direct sunlight.
   (b) Microclimate.

3. (a) Rock pools may have low salinity due to rain falling into them directly or through runoff.
   (b) Rock pools may have very high salinity due to evaporation after long exposure times without rain.

4. Environmental gradients from canopy to leaf litter:
   (a) Light intensity: decreases.
   (b) Wind speed: decreases.
   (c) Humidity: increases.

5. Reasons why factors change:
   (a) Light intensity: Foliage above will shade plants below, with a cumulative effect. The forest floor receives light that has been reflected off leaf surfaces several times, or passed through leaves.
   (b) Wind speed: Canopy trees act as a wind-break, reducing wind velocity. Subcanopy trees will reduce the velocity even further, until near the ground the wind may be almost non-existent. An opening in the forest canopy (a clearing) can expose the interior of the forest to higher wind velocities.
   (c) Humidity: The sources of humidity (water vapour) are the soil moisture, leaf litter, and the transpiration from plants. Near the canopy, the wind will carry away moisture-laden air. Near the forest floor, there is little wind, and humidity levels are high.

6. The colour of the light will change nearer the forest floor. White light (all wavelengths) falling on the canopy will be absorbed by the leaves. Reflected light in the green wavelength bounces off the leaves and passes downward to lower foliage and the forest floor.

7. (a) Plants growing close to the forest floor are less susceptible to wind and water loss.
   (b) Plants growing closer to the forest floor have a lower amount of available light.

8. Environmental gradients from water surface to bottom:
   (a) Water temperature: Decreases gradually until below the zone of mixing when there is a sharp drop.
   (b) Dissolved O_2: Oxygen at a uniform concentration until below the zone of mixing when there is a sharp drop, with very little oxygen at the bottom.
   (c) Light penetration: Decreases at an exponential rate (most light is absorbed near the surface).

9. (a) Prevents mixing of the oxygen-rich surface water with the deeper oxygen-deficient water (represents a thermal barrier).
   (b) Organisms (particularly bacteria) living below the thermocline use up much of the available oxygen. Decomposition also uses up oxygen.

10. (a) Heavy rainfall or inflow of floodwater from nearby river channels may cause a decline in salinity.
    (b) Evaporation from the lake concentrates salts and the conductivity will increase.

11. Physical gradients will govern what species will be found and where in a particular area, as determined by their specific tolerances to abiotic factors.

123. Habitat (page 166)

1. An organism will occupy habitat according to its range of tolerance for a particular suite of conditions (temperature, vegetation and cover, pH, conductivity). Organisms will tend to occupy those regions where all or most of their requirements are met and will avoid those regions where they are not. Sometimes, a single factor, e.g. pH for an aquatic organisms, will limit occupation of an otherwise suitable habitat.

© 2016 **BIOZONE** International
ISBN: 978-1-927309-22-3
Photocopying Prohibited

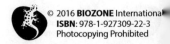

2. (a) Most of a species population is found in the optimum range because this is where conditions for that species are best; most of the population will select that zone.
   (b) The greatest constraint on an organism's growth within its optimum range would be competition between it and members of the same species (or perhaps different species with similar niche requirements).

3. In a marginal niche, the following might apply:
   – Physicochemical conditions (e.g. temperature, current speed, pH, salinity) might be sub-optimal and create stress (therefore greater vulnerability to disease).
   – Food might be more scarce or of lower quality/nutritional value.
   – Mates might be harder to find.
   – The area might be more exposed to predators.
   – Resting, sleeping, or nesting places might be harder to find and/or less suitable in terms of shelter or safety.
   – Competition from other better-adapted species might be more intense.

124. Ecological Niche (page 167)

1. (a) The realised niche of a species can be narrower or broader depending on the constraints that other species place on physical space and resource use.
   (b) Competition will exert the greatest constraint on the extent of an organism's niche. To a lesser extent, so too will parasitism, predation, and disease.

2. Interspecific competition will result in some overlap in resource use curves and selection will favour a contraction of the niche and species specialisation (of resource use). The (favourable) result of this is a divergence in resource use curves. Intraspecific competition in contrast, acts to broaden niches because competing individuals are forced to exploit resources at the extremes of their tolerance range. Because the competing individuals are the same species (with the same resource requirements) specialisation into different niches is generally not an option (within the constraints of the species biology) so intraspecific competition is a major constraint on population size. **Teacher's explanatory note**: Of course, there are a number of examples of ecologically and genotypically flexible species moving into new regions unoccupied by competitors. In these cases, character displacement within one species can result in speciation given the appropriate conditions of isolation and resource and niche availability. The finches of the Galàpagos are one example. In most ecological situations though, intense competition within a species when niches are limited forces a broadening of that species niche.

125. Factors Determining Population Growth (page 168)

1. (a) Mortality: Number of individuals dying per unit time (death rate).
   (b) Natality: Number of individuals born per unit time (birth rate).
   (c) Net migration rate: Net change in population size per unit time due to immigration and emigration.

2. Limiting factors place a constraint on population growth, abundance, or distribution. The constraint acts thorough the scarcest necessary resource not the total available resources.

3. (b) Declining population: $B + I < D + E$
   (c) Increasing population: $B + I > D + E$

4. (a) Birth rate = 14 births ÷ 100 total number of individuals x 100 % = 14% per year
   (b) Net migration rate = 2% per year
   (c) Death rate = 20% per year
   (d) Rate of population change: birth rate – death rate + net migration rate = $14 - 20 + 2 = -4\%$ per year
   (e) The population is declining.

5. Availability of food and availability of water.

126. Population Growth Curves (page 169)

1. As the population increases, it encounters environmental resistance to further growth. In short, they run out of resources and reach carrying capacity.

2. Environmental resistance refers to all the limiting factors that together act to prevent further population increase (achievement of intrinsic rate of population increase, r_{max}).

3. (a) The maximum population size (of a species) that can be supported by the environment.
   (b) Carrying capacity limits population growth because as the population size increases, population growth slows (when $N = K$ population growth stops).

4. (a) Early in its growth, the population grows at an exponential rate. At the transitional phase, the population encounters resistance to exponential growth, known as environmental resistance. The rate of population increase slows and stabilises around carrying capacity.
   (b) The population response (in numbers) that we see is the sum result of the preceding birth and death rates, so populations lag in their response to environmental change and may overshoot K before lack of resources causes a decline in population growth rate.

5. Students can use the spreadsheet on the Teacher's Digital Edition and the weblinks examples if they need help.

127. Plotting Bacterial Growth (page 170)

1. Completed table below:

Min	No.	Min	No.	Min	No.
0	1	140	128	260	8 192
20	2	160	256	280	16 384
40	4	180	512	300	32 768
60	8	200	1024	320	65 536
80	16	220	2048	340	131 072
100	32	240	4096	360	262 144
120	64				

2. (a) 8   (b) 512   (c) 262 144

3. and 4(b)

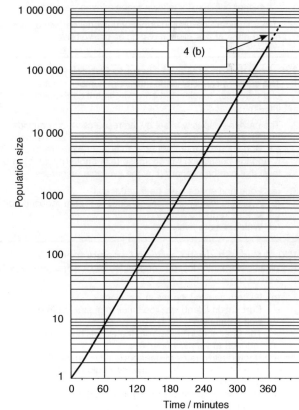

4. (a) 524 288
   (b) See graph above

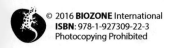

5. With exponential growth, the numbers are very low initially, but increase quickly and very large numbers are involved. The log graph makes it feasible to plot the very small and very large numbers together in a reasonable space and in a way that is easy to read and interpret.

128. Investigating Bacterial Growth (page 171)

1. The precautions prevent the accidental introduction of the bacteria into the environment and prevent accidental infection.

2. (a) See graph below.

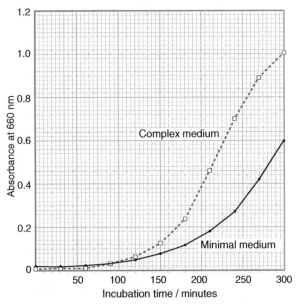

(b) The absorbance measures the amount of light absorbed by the sample (more light absorbed means a higher density of cells).
(c) The complex medium supported more rapid growth to a higher cell density relative to the minimal medium.

3. Use a standard calibration curve of bacterial density against absorbance or use dilution plating as per weblink.

129. Population Size and Carrying Capacity (page 172)

1. The carrying capacity is the maximum number of individuals of a particular species that an environment can support indefinitely.

2. The carrying capacity is set by the resources it can provide and these are limited. If a population increases above the carrying capacity, there will be insufficient resources to sustain it, and the population will decrease (e.g. through deaths) to a level that can be supported by the available resources.

3. (a) Food and space have been reduced.
(b) Available water (and consequently food) are reduced due to the drought.
(c) Water is more available.

130. A Case Study in Carrying Capacity (page 173)

1. Wolves were introduced to control the black-tailed deer, which were overgrazing the land.

2. (a) Factors causing the result included:
    –   Coronation Island was too small to sustain both deer and wolf populations.
    –   The deer couldn't hide from the wolves so could be reduced to very low numbers.
    –   Reproductive rates of deer could have been low because of poor forage, so the population could not withstand predation.
    –   There were no other prey so no opportunity for prey switching when deer became scarce.
(b) The carrying capacity of Coronation Island is too low to support viable (sustainable) populations of a large predator (wolf) and its prey (deer).

131. Species Interactions (page 174)
Erratum: Commensalism (one species benefits and the other is unaffected) has been defined in the second printing of AQA 2.

1. (a) Mutualism: Domesticated animals (e.g. dogs and cats in western culture, work horses) and plants not grown for consumption.
(b) Exploitation: Using plants and animals for food source, skins/pelts for clothing, timber and other plant products for shelter and building materials.
(c) Competition: Invertebrate pests and some fungi feeding on our crops (e.g. insects such as aphids, locusts, caterpillars; slugs, snails, mildew, rusts).

2. (a) Acacia produces toxic alkaloids in response to browsing.
(b) This response makes the giraffe move on to another plant before it removes too much of the acacia's foliage.

3. (a) Mutualism
(b) Both species benefit. Flowers are pollinated and bees gain food.

4. (a) Commensalism or perhaps mutualism
(b) The anemone shrimp benefits by gaining protection while the anemone is (apparently) unaffected. However, there is some limited evidence that anemones may benefit although this isn't confirmed. The anemones house photosynthetic algae in their tissues that produce nutrients that are used by the anemone. These algae benefit from the ammonium released by the shrimps so indirectly, the anemone also benefits.

5. (a) Predation
(b) The hyaena benefits while the prey is harmed (killed).

6. (a) Parasitism
(b) The tick benefits while the host is harmed.

7. (a) Commensalism or mutualism
(b) The egret benefits by feeding on the insects disturbed by the herbivore. The herbivore is either not affected or gains a small benefit by having annoying (and potentially disease-carrying) insects removed from the grazing area.

8. Both feed off another organism but, unlike a predator, a parasite does not kill its host.

132. Interpreting Predator-Prey Relationships (page 176)

1. (a) Peak numbers of woolly aphids and ladybird marked on graph (below).

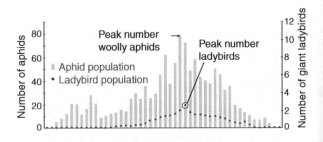

(b) No, they are slightly offset (ladybirds lag).
(c) Giant ladybirds feed only on woolly aphids, so ladybird numbers can only increase if there is enough food (woolly aphids) to sustain population growth. Giant ladybird numbers will respond to the woolly aphid numbers and slightly lag behind.

2. (a) Positive.
(b) Giant ladybird numbers follow the trend for woolly aphid numbers. When woolly aphid numbers are increasing, the giant ladybird numbers increase. As woolly aphid numbers decrease so do the giant ladybird numbers.

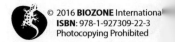
© 2016 **BIOZONE** International
ISBN: 978-1-927309-22-3
Photocopying Prohibited

133. Interspecific Competition (page 177)

1. The two species have similar niche requirements (similar habitats and foods). Red squirrels once occupied a much larger range than currently. This range has contracted steadily since the introduction of the greys. The circumstantial evidence points to the reds being displaced by the greys.

2. The greys have not completely displaced the reds. In areas of suitable coniferous habitat, the reds have maintained their numbers. In some places the two species coexist. **Note**: It has been suggested that the reds are primarily coniferous dwellers and extended their range into deciduous woodland habitat in the absence of competition.

3. Habitat management allows more effective long term population management *in-situ* (preferable because the genetic diversity of species is generally maintained better in the wild). Reds clearly can hold their own in competition with greys, provided they have sufficient resources. **Enhancing the habitat** preferred by the reds (through preservation and tree planting), assists their success. Providing **extra suitable food plants** also enables the reds to increase their breeding success and maintain their weight through winter (thus entering the breeding season in better condition).

4. Other conservation strategies to aid red squirrel populations could include (any of): Captive breeding and release of reds into areas where they have been displaced, control/cull of grey squirrels (particularly in habitats suitable for reds), transfer of reds from regions where populations are successful to other regions of suitable habitat, supplementary feeding prior to the breeding season, public education to encourage red squirrels over greys.

5. (a) A represents the **realised niche** of *Chthamalus*.
   (b) When *Balanus* is removed from the lower shore, the range of *Chthalamus* extends into areas previously occupied by the *Balanus*; *Balanus* normally excludes Chthamalus from the lower shore.

134. Niche Differentiation (page 179)

1. (a) When resources are limited during the winter, coal tits forage in the needles and upper parts of the tree, goldcrests forage near the ground, crested tits forage in the lower parts of the tree, and blue tits forage in the middle heights of pine trees.
   (b) When other birds are absent, coal tits occupy more of the tree than when other species are present. In addition, when food resources are more abundant, the various tit species forage in more parts of the tree than in winter and more of the foraging locations overlap.

2. Generally, the foraging ranges of the birds overlap during warmer months when food is more plentiful but they restrict their foraging to separate regions of the tree in winter when food is scarce. When resources are more abundant, coal tits occupy more of the tree and spend more time foraging in the needles. In winter, they forage more in the upper parts of the tree. Goldcrests forage near the ground in winter but higher up the tree in warmer months. Crested tits spend much more time foraging in the lower parts of the tree during winter but will forage in every part of the tree (overlapping with other species) in warmer months. Blue tits forage in the middle parts of the tree but feed higher in the tree in winter than in summer.

135. Intraspecific Competition (page 180)

1. (a) Individual growth rate: Intraspecific competition may reduce individual growth rate when there are insufficient resources for all individuals. Examples: tadpoles, *Daphnia*, many mammals with large litters. Explanatory note: Individuals compete for limited resources and growth is limited in those that do not get access to sufficient food.
   (b) Population growth rate: Intraspecific competition reduces population growth rate. Examples as above. Explanatory note: Competition intensifies with increasing population size and, at carrying capacity, the rate of population increase slows to zero.
   (c) Final population size: Intraspecific competition will limit population size to a level that can be supported by the carrying capacity of the environment. Note: In territorial

species, this will be determined by the number of suitable territories that can be supported.

2. (a) They reduce their individual growth rate and take longer to reach the size for metamorphosis.
   (b) Density dependent.
   (c) The results of this tank experiment are unlikely to represent a real situation in that the tank tadpoles are not subject to normal sources of mortality and there is no indication of long term survivability (of the growth retarded tadpoles). Note: At high densities, many tadpoles would fail to reproduce and this would naturally limit population growth (and size) in the longer term.

3. Reduce intensity of intraspecific competition by:
   (a) Establishing hierarchies within a social group to give orderly access to resources.
   (b) Establishing territories to defend the resource within a specified area.

4. (a) Carrying capacity might decline as a result of unfavourable climatic events (drought, flood etc.) or loss of a major primary producer (plant species).
   (b) Final population size would be smaller (relative to what it was when carrying capacity was higher).

5. Territoriality is a common consequence of intraspecific competition in mammals and birds. In any habitat, resources are limited and only those with sufficient resources will be able to breed. This is especially the case with mammals and birds, where the costs of reproduction to the individual are high relative to some other taxa. Even though energy must be used in establishing and maintaining a territory, territoriality is energy efficient in the longer term because it gives the breeding pair relatively unchallenged access to resources. As is shown in the territory maps of golden eagles and great tits, territories space individuals apart and reduce intraspecific interactions. The size of the territory is related to the resources available within the defended area; larger territories are required when resources are poorer or widely dispersed. As is shown by the great tit example, when territory owners are removed, their areas are quickly occupied by birds previously displaced by competition.

136. Quadrat Sampling (page 182)

1. Mean number of centipedes captured per quadrat:
   Total number centipedes ÷ total number quadrats
   30 individuals ÷ 37 quadrats
   = 0.811 centipedes per quadrat

2. Number per quadrat ÷ area of each quadrat
   $0.811 ÷ 0.08 = 10.1$ centipedes per m^2

3. Clumped or random distribution.

4. Presence of suitable microhabitats for cover (e.g. logs, stones, leaf litter) may be scattered.

137. Quadrat-Based Estimates (page 183)

1. Species abundance in plant communities can be determined by using quadrats and transects, and abundance scales and percentage cover are often appropriate. Methods for sampling animal communities are more diverse, and density is a more common measure of abundance.

2. **Size**: Quadrat must be large enough to be representative and small enough to minimise the amount of sampling effort.

3. **Habitat heterogeneity**: Diverse habitats require more samples to be representative because they are not homogeneous.

4. (a) and (b) any two of:
   - The values assigned to species on the abundance scale are subjective and may not be consistent between users.
   - An abundance scale may miss rarer species and overestimate conspicuous ones.
   - The scale may be inappropriate for some habitats.
   - The semi-quantitative values assigned to the abundance categories cover a range so results will lack precision.

138. Sampling a Rocky Shore Community (page 184)

The results per se are not particularly important, but it is important to understand the method and its limitations. The results will vary depending on a group's agreed criteria for inclusion of organisms in a given quadrat (e.g. when and how an organism is counted when it is partly inside a quadrat). **Note:** Some algae are almost obscured by organisms or have other algae on top of them.

6. Typical results (total for each category) are:

	A	B	C	D	Direct count
Barnacle	9	4	10	13	81
Oyster borer	0	0	1	1	3
Chiton	1	0	1	0	3
Limpet	0	3	0	0	6
Algae	24	13	13	14	103

7. Typical results for calculated population density based on A-D and on a direct count (question 8b):

	A	B	C	D	Direct count
Barnacle	1667	741	1852	2407	2500
Oyster borer	0	0	185	185	93
Chiton	185	0	185	0	93
Limpet	0	556	0	0	185
Algae	4444	2407	2407	2593	3179

Note: Area of 6 quadrats = (0.03 x 0.03) x 6 = 0.0054 m^2
Area of total sample area = 0.18 x 0.18 = 0.0324 m^2

8. (a) Once the quadrats have been laid, the animals moving from one quadrat to another might be counted twice. The quadrat could involve the placement of physical barriers between each quadrat. There is a possibility of exposing the entire area and photographing it for later analysis.
   (b) Densities calculated on direct count in the last column of the table above. Students should be aware of the dangers of extrapolating data from a small sample. Including or excluding single individuals can have a large effect on the density calculated, particularly where species are present at low densities. Extension: Groups could combine data to see if they get a more representative sample (i.e. closer to the direct count).

139. Transect Sampling (page 186)

1. (a) With belt transects of 10 m or more, sampling and analysis using this method is very time consuming and labour intensive.
   (b) Line transects may not be representative of the community. There may be species which are present but which do not touch the line and are not recorded.
   (c) Belt transects use a wider strip along the study area so there is less chance that a species will not be recorded.
   (d) Situations involving highly mobile organisms.

2. Decrease the sampling interval. If no more species are detected and the trends along the transect remain the same, then the sampling interval was adequate.

3.

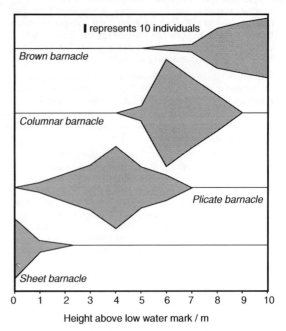

4. (a) Percentage cover of each seaweed species.
   (b) Seaweed vigour and degree of dessication.

140. Qualitative Practical Work: Seaweed Zonation (page 188)

1. (a) Percentage cover of each seaweed species.
   (b) Seaweed vigour and degree of dessication.

2. Column graph as below:

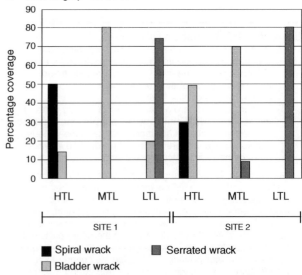

3. Spiral wrack is the most tolerant of long exposure times, where it grows vigorously despite showing some evidence of desiccation. Bladder wrack grows vigorously throughout the midlittoral and is relatively tolerant of exposure, only showing signs of desiccation higher on the shore where exposure times are longer. Serrated wrack is intolerant of exposure and grows vigorously at the LTL but shows signs of desiccation above this and cannot compete in the midlittoral zone with the more tolerant bladder wrack.

4. Quadrat position was staggered for the two sites to give a better indication of the extent of each species' distribution. The disadvantage is that the sites cannot be directly compared.

141. Mark and Recapture Sampling (page 189)

1. Results will vary from group to group for this practical. The actual results are not important, but it should serve as a useful vehicle for discussion of such things as sample size, variation in results between groups, and whether the method

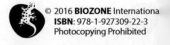
© 2016 **BIOZONE** International
ISBN: 978-1-927309-22-3
Photocopying Prohibited

is a reliable way to estimate the size of a larger unknown group. Discussion could centre around what factors could be altered to make it a more reliable method (e.g. larger sample size, degree of mixing, increasing number of samples taken).

2. Trout in Norwegian lake:
   Size of 1st sample: 109
   Size of 2nd sample: 177
   No. marked in 2nd sample: 57
   Estimated total population: 109 x 177 ÷ 57 = 338.5

3. (a) Some marked animals may die.
   (b) Not enough time for thorough mixing of marked and unmarked animals.

4. (a) and (b) in any order (any two of):
   – Marking does not affect their survival.
   – Marked & unmarked animals are captured randomly.
   – Marks are not lost.
   – The animals are not territorial (must mix back into the population after release).

5. (a) Any animal that cannot move or is highly territorial (e.g. barnacle, tube worm, many mammals).
   (b) Unable to mix with unmarked portion of the population. Recapture at the same location would simply sample the same animals again.

6. (a)-(c) in any order:
   – Banding: leg bands of different colour on birds.
   – Tags: crayfish shell, fish skin, mammal ears.
   – Paint/dye used to paint markings in shell/fur.

7. The scientists obtain information on fish growth to establish the relationship between age and growth. This will help manage the population to prevent overfishing. Tracking also helps to map breeding grounds and migrations so that fish can be protected at critical times in their life histories. In addition to these data, researchers will find out more about the general biology of the cod (e.g. data on feeding), which will help in the management and recovery of the fish stock.

142. Field Study of a Rocky Shore (page 191)

1. Hypothesis (c): The physical conditions of exposed rocky shores and sheltered rocky shores are very different and so the intertidal communities will also be different.

2. See table below.

3. See graph below.

4. (a) The mean, medians, and modes are all similar.
   (b) The data are normally distributed.

5. (a) Rock oyster
   (b) Site A is open to the swell, which dislodges the oysters. Site B is more sheltered.

6. (a) Brown and plicate barnacles have a preference for exposed rocky shores.
   (b) Oyster borers are predators of brown and plicate barnacles so are more abundant when brown and plicate barnacles are also abundant.

7. (a) There are relatively high numbers of limpets at each site.
   (b) Limpets have a wide range of tolerance and are therefore able to live in many areas.

Table: Total and mean numbers of intertidal animals at two rocky shore sites.

		Brown barnacle	Oyster borer	Columnar barnacle	Plicate barnacle	Rock oyster	Ornate limpet	Radiate limpet	Black nerite
Site A	Total number of animals	308	46	78	386	0	63	47	55
	Mean animals per m²	39	6	10	48	0	8	6	7
	Median value	38.5	6	9.5	51	0	7.5	6	7
	Modal value	-	7	8	-	-	-	6	-
Site B	Total number of animals	52	15	427	85	49	50	96	29
	Mean animals per m²	7	2	53	11	6	6	12	4
	Median value	7	2	56.5	10.5	7.5	6.5	12.5	3.5
	Modal value	7	-	58	-	8	5	14	-

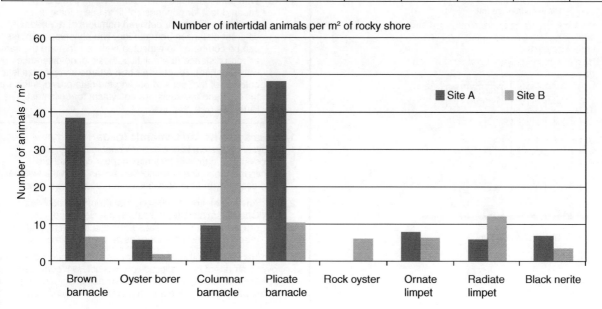

Number of intertidal animals per m² of rocky shore

143. Investigating Distribution and Abundance
(page 194)

1.

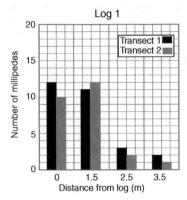

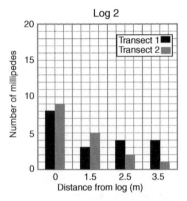

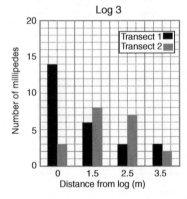

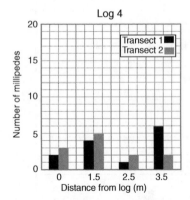

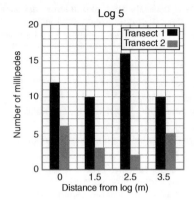

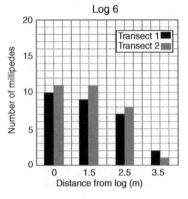

2. (a) Yes. In most cases, there are a larger number of millipedes closer to the log.
   (b) It is likely that the nutrient and moisture levels are higher close to the rotting log, which would form a microclimate.
   (c) Physical conditions such as temperature, light levels, or relative humidity could be measured. This would test the assumption that nutrient and moisture levels are higher close to the log and allow comparison of the physical conditions of equivalent transects each side of the log.

3. (a) There are valid reasons for saying either yes or no.
   (b) For no: The data from the study show that in some cases, but not all, the left side of the log (transect 1) generally has more millipedes than the right side (transect 2).
   For yes: Logs 1, 2, and 4 are close to even or even. Log 6 has greater numbers on the right than the left and 3 and 5 have more on the left than the right. The differences are not consistent and may not be statistically significant.

4. (a) With pooled data at each distance you could a chi squared test for goodness of fit to see if there is a significant difference between numbers of millipedes at different distances from the log. The observed numbers can be compared to expected numbers that are the same at each distance (the null hypothesis of no difference with distance). **Teacher's note**: A large sample size is a criterion of the test and pooling the data provides this (as the logs and transects are equivalent to replicates).
   (b) Student's own response.

144. Ecosystems Are Dynamic (page 196)

1. A dynamic system is one that is constantly changing. Ecosystems, although they may appear constant are continually changing in response to changes in the weather and seasons and the activities of the organisms in them.

2. (a) Any two of: fire, flood, seasonal drought, landslides.
   (b) Open pit mining, large scale forest clearance, volcanic eruptions, inundation caused by sea level rise, prolonged drought (desert formation) as caused by climate shifts.

3. A climax community is one that apparently remains the same over time (there is no succession to a different community). However, an ecosystem is constantly adjusting to changes in the physical environment. In a climax community, an

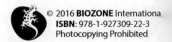

© 2016 **BIOZONE** International
ISBN: 978-1-927309-22-3
Photocopying Prohibited

equilibrium exists between growth and death in the community so that it appears static over the long term.

145. Ecosystems are Resilient (page 197)

1. Ecosystem resilience is the ability of the ecosystem to recover after a disturbance e.g. fire.

2. Middle reef experienced fewer disturbances giving time for coral to regain its dominance, whereas frequent disturbances at Low Isles make it very difficult for the coral to grow and remain dominant there.

3. The input of energy from the farmer in the form of weeding, sprays, and pest eradication. This will only maintain the monoculture until the plants mature and are harvested unless there is continual replanting.

146. Ecosystem Changes (page 198)

1. (a) When dry, peat is used as a fuel source.
   (b) Peat forms very slowly and removing it destroys the system and the conditions that maintained it. It may take hundreds or thousands of years for a mined peat bog to recover (if ever).
   (c) Mining below the water table creates a lake (open water) ecosystem, which are vastly different to the peat bog that was there before. The lake removes any chance of the peat bog recovering and also meant the land was no longer available for agriculture.

2. Loss of the fir, cedar, and pine forests, which were replaced by volcanic ash.

3. This area was closest to the volcanic blast and consequently suffered the largest amount of damage. Its altitude and covering of volcanic ash make it difficult for plants to reestablish there.

4. Coldwater Creek was blocked, forming Coldwater lake, changing the system from flowing water to static water.

5. Spirit lake was completely emptied and all life, except bacteria, was killed. When the refilled lake recovers, its new community will reestablish under quite different conditions and its characteristics are likely to be quite different.

6. (a) The river now originates at the crater instead of Spirit Lake and it is now heavily laden with sediment.
   (b) The original community was adapted to a clear water environment. The river is now laden with sediment and the community will change to one that is tolerant of the different conditions. Species that depend on clear water will disappear.

147. Primary Succession (page 200)

1. Glacial retreat, exposed slip, new volcanic island.

2. (a) Lichens, mosses, liverworts, and hardy annual herb species are often the first to colonise bare ground.
   (b) (i) Chemically and physically erode rock (producing the beginnings of a soil). (ii) Add nutrients by decay.

3. Climax communities tend to have higher biodiversity and a more complex trophic structure than early successional communities. A greater diversity of community interactions buffers the system against disturbances because there are many more organisms with different ecological roles able to compensate for losses from the system.

148. Succession on Surtsey Island (page 201)

1. Surtsey was ideal as a study site for primary succession because it was an entirely new island, devoid of any soil, and was isolated from nearby influences (such as already established vegetation or urban settlements) that could accelerate the succession process.

2. Early colonisations were the result of seeds blown from Iceland to the Northern shore, which is the closest shore. Later colonisations were in the south due to the establishment of the gull colony. The gulls would transport seeds and contribute to soil fertility.

3. (a) 1985.
   (b) Transported by birds.

(c) 1985. This coincides with the establishment of the gull colony as the gulls were instrumental in dispersing seeds.

149. Conservation and Succession (page 202)

1. Conservation projects sometimes need to slow or stop succession to maintain the habitat on which the species of interest rely. Normal successional changes to the environment might alter the habitat enough to make it difficult for the species there to survive. This is particularly true of habitats that have been maintained in a plagioclimax state through management for hundreds of years and have a unique flora and fauna that is worthy of conserving.

2. Actively managing ecosystems to deflect the normal successional process helps to maintain the existing biodiversity of the plagioclimax community. For example, grazing and burning of moorland maintains the diversity of moorland vegetation and prevents the establishment of trees (and invasion of undesirable plants such as bracken).

150. Conservation and Sustainability (page 203)

1. Conservation refers to the management of a resource so that it is maintained into the future. The resource may be a living or mineral system and may focus on the efficient use of a non-renewable resource or the managed use of a living system. Sustainability refers to the idea of using resources within the capacity of the environment and the eventual replacement of what has been used. It can be viewed as a subset of conservation, mainly focussing on living systems.

2. Resource conservation is the efficient use of resources so that stocks remain available into the future. This means not wasting non-renewable resources (e.g. oil) and making sure renewable resources can be replenished at the same (or greater) rate as they are used (e.g. fish stocks).

3. Question should read: What is the importance of society and economy **in** the conservation of living systems and resources? Society and the economy are important in conservation because a strong economy and a stable, equitable society contribute to higher employment, better living standards, and a higher level of education. These in turn promote sound infrastructure (e.g. sewage disposal), increased awareness of the environment and resource use, and economic decisions that support long term conservation goals. If society is disrupted, e.g. by internal or external conflict, and the economy is weak, resources will be exploited, perhaps unsustainably, just to survive.

151. Sustainable Forestry (page 204)

1. (a) Coppicing is a low-impact method of producing wood for harvest, which preserves a rich diversity of woodland species and is sustainable in the long term. However, it is a low volume method and the timber is not suitable for all uses. It is also skilled and labour intensive work and, if not carried out properly, has detrimental effects on diversity and forest structure.
   (b) Strip cutting harvests a forest in strips and has the advantage that it produces relatively high volumes of timber suitable for a wide range of uses while leaving enough border forest to enable rapid recovery of cleared land. It is also a low effort for return method and has a relatively low impact on the environment and on biodiversity. Most disadvantages come from the infrastructure required to carry out the forestry (roads etc) and the time it takes for the forest to regenerate (so for slow growing species it is not suitable).
   (c) Clear cutting is a high volume method with significant detrimental impacts on the environment (e.g. high soil erosion, habitat loss, reduced biodiversity). The forest requires replanting and, if suitable at all, clear cutting should be restricted to fast growing plantation forests.
   (d) Selection logging is a medium volume supplier of timber, but causes a moderate amount of environmental damage, which depends partly on how much infrastructure and machinery are associated with the timber extraction. Effects on biodiversity are usually minimal, and can be sustainable if properly managed to retain the original forest composition.

2. (a) Commercial plantations: Clear cutting.
   (b) Traditional woodland: Coppicing.
   (c) Second growth: Strip cutting or selective logging.

3. The student's own opinion is important here and should be qualified. Given unlimited resources, coppicing provides long term sustainability, with volume supplied from a mix of selective logging and strip cutting in second growth and plantation forests. Logging of old growth forests should never be considered sustainable.

4. Properly coppiced woodlands provide a regular timber supply and maintain an open canopy that encourages a rich understorey and faunal diversity. If a previously coppiced woodland is allowed to deteriorate without management, it becomes tall and overshaded, yet lacking the characteristics of an old growth forest. Biodiversity declines markedly. These are strong arguments for managing coppiced lands as a long term resource.

152. Sustainable Fishing (page 206)

1. Without accurate estimates of age, population size and growth rate it is impossible to calculate an accurate MSY. Incorrect calculation of the MSY may lead to over fishing and the collapse of the fishery, or to under fishing with not enough fish being landed to create a viable economic resource.

2. (a) Overestimating the population size leads to overestimating replenishment rate and the size of the sustainable catch.
   (b) Overestimating the growth rate leads to overestimating the size of the catch that can be landed due to the assumption the population will quickly recover its loses.
   (c) This scenario could lead to the MSY being set too low due to the assumption that the population ages and replaces its losses slowly. It may also lead to the assumption that the population is close to collapse.

3. The statement is correct in that the population can only be harvested at the MSY if the population growth rate remains stable. If the rate drops but the MSY remains the same then each season a greater proportion of the population will be taken. For example, a population of 100 has 40 taken each season (leaving 60). If a bad season sees the population recover to only 90 before the next harvest of 40 the population will be reduced to 50. A second bad season may see a further reduction. Harvesting at the MSY leaves the population vulnerable to changes in its population growth. Because of this, allowed fishing take is normally set below the MSY.

4. (a) Total catch has declined and is now nearly 300 000 tonnes less than during peak harvests in the 1970s. Mortality of mature fish increased to a peak around 1990-2000 but has declined steadily since then. Spawning and recruitment rates have both declined dramatically, although spawning stock biomass has recovered slightly since 2005.
   (b) There will be little effect on stock recovery. This measure will only slow its decline, rather than increase numbers. This is because the current estimated maximum mortality rate (due to fishing) is 0.2, which is lower than the fisheries target of 0.4.

153. Modelling a Solution (page 208)

1. This activity is based entirely on the student's ideas and justifications. There is therefore no right or wrong answer.

154. Chapter Review (page 209)

No model answer. Summary is the student's own.

155. KEY TERMS AND IDEAS: Did You Get It? (page 211)

1. Alleles (D), autosome (G), diploid (A), dominant (I), genotype (F) monohybrid cross (C), phenotype (E), Punnett square (H), recessive (B).

2. (a) Pea seeds with the genotype YyRr would be yellow-round.
   (b) Punnett square:

	YR	Yr	yR	yr
YR	YYRR	YYRr	YyRR	YyRr
Yr	YYRr	YYrr	YyRr	Yyrr
yR	YyRR	YyRr	yyRR	yyRr
yr	YyRr	Yyrr	yyRr	yyrr

Yellow-round: 9/16   Yellow-wrinkled: 3/16
Green-round: 3/16   Green-wrinkled: 1/16

3. Directional selection describes the situation where a phenotype at one extreme of the phenotypic range has the greatest fitness, so the phenotypic norm shifts in the direction of that phenotype, e.g. selection for dark morphs of *Biston* moths in England during the Industrial Revolution. Stabilising selection describes the situation where fitness is highest for the most common phenotype and there is selection against phenotypes at the extremes of the phenotypic range, e.g. selection for human birth weight.

4. Allopatric speciation (H), conservation (N), founder effect (I), gene pool (C), genetic bottleneck (B), genetic drift (L), Hardy-Weinberg principle (J), interspecific competition (D), intraspecific competition (G), mark and recapture (M), natural selection (A), primary succession (P), quadrat (O), reproductive isolation (K), speciation (E), sympatric speciation (F).

5. (a) Approximately 6000 individuals.
   (b) Population was entering a phase of exponential growth.
   (c) The population had crashed to a low point.
   (d) The population had exceeded carrying capacity by a large amount and would have run out of resources. Many individuals would have died and population growth would have fallen dramatically.

156. What is a Gene Mutation (page 215)

1. Mutations that alter the amino acid sequence are often harmful because they result in a non-functional protein.

2. A missense substitution results in a polypeptide chain with an incorrect amino acid which may have little or no effect on the protein produced (due to code degeneracy and depending on where the substitution occurs. In contrast, a nonsense substitution produces a codon that doesn't code for an amino acid or a STOP codon which prematurely ends synthesis of the polypeptide chain.

3. (a) A silent mutation is a point mutation that does not result in a change in the amino acid sequence (because of the degeneracy of the genetic code).
   (b) Silent mutations are considered neutral if they do not alter fitness (do not change selection pressures).

157. Reading Frame Shifts (page 216)

1. A reading frame shift is an alteration in the order in which the 3 base sequence (codon) is read occurring as a result of a base insertion or deletion (reading frame is offset). Reading frame shifts produce an entirely new amino acid sequence downstream of the mutation.

2. A frame shift near the start codon will have a greater impact than one near the stop codon because more of the codons (reading frame) are affected, i.e. the base offset occurs earlier and more amino acids in the sequence will be incorrect.

© 2016 **BIOZONE** Internationa
ISBN: 978-1-927309-22-3
Photocopying Prohibited

3. An altered chain may or may not produce a protein with biological activity depending on whether the frame shift occurs in a critical part of the protein, e.g. in the critical amino acid interactions that create the active site of an enzyme.

158. Chromosome Mutations (page 217)

1. (b) Original sequence: ABCDEFGHMNOPQRST
        Mutated sequence: ABFEDCGHMNOPQRST
    (c) Original sequences: 1234567890
                             ABCDEFGHMNOPQRST
        Mutated sequences: ABCDEF1234567890
                             GHMNOPQRST
    (d) Original sequences: ABCDEFMNOPQ
                             ABCDEFMNOPQ
        Mutated sequences: ABCDEABCDEFMNOPQ
                             FMNOPQ

2. Inversion, because there is no immediate loss of genes from the chromosome. (At a later time, inverted genes may be lost from a chromosome during crossing over, due to unequal exchange of segments).

159. Sickle Cell Mutation (page 218)

1. (a) Bases: 21   (b)   Triplets: 7   (c) Amino acids: 7

2. (a) A point mutation, a base substitution, causes one amino acid to change in the β-chain of the haemoglobin molecule.
    (b) The mutated haemoglobin protein behaves differently to the normal haemoglobin, so that when not carrying oxygen, it precipitates out into fibres which deform the RBC to a sickle cell shape.
    (c) Heterozygotes (carriers) of the sickle cell mutation have both normal and mutated haemoglobin molecules. They have enough functional haemoglobin to carry sufficient oxygen, so only suffer minor effects of the disease.
    (d) The mutated haemoglobin is less soluble at low oxygen tensions. In low oxygen environments (such as at altitude) the mutated haemoglobin will precipitate out and the carrier will show symptoms of sickle cell disease.

3. The sickle cell mutation affords some degree of resistance to malaria and so persists where malaria is present.

160. Cystic Fibrosis Mutation (page 219)

1. (a) mRNA: GGC ACC AUU AAA GAA AAU AUC AUC UUU GGU GGU
    (b) Mutant mRNA: GGC ACC AUU AAA GAA AAU AUC AUC I GGU GGU
    (c) A triplet deletion.

2. (a) The abnormal CFTR protein is rapidly degraded in the cell and so does not insert into the plasma membrane.
    (b) Water moves into the cells because of the accumulation of chloride ions.

161. What Are Stem Cells? (page 220)

1. (a) Potency - ability to differentiate into other cell types.
    (b) Self renewal - ability to maintain an unspecialised state.

2. The differentiation of cells into specialised types is the result of specific patterns of gene expression and is controlled by transcription factors in response to cues (e.g. from hormones) during development.

3. Stem cells differentiate to give rise to all the cell types of the multicellular organism. The zygote is totipotent and can differentiate into any cell type, including extra-embryonic cells. As development proceeds, the cells become pluripotent. They can give rise to most cell types, but not extra-embryonic cells, and are important in the early development of tissues and organs. Throughout life, multipotent adult stem cells maintain these tissues and organs and are committed to produce the different cell types related to the tissue of origin.

4. (a) Pluripotent stem cells are naturally found in the inner mass of the blastocyst and can give rise to any cells of the body (except extra-embryonic cells). iPSC are induced from somatic cells by the insertion of specific genes and give rise to unipotent stem cells which give rise only to one specific cell type.

    (b) The insertion of genes, known as reprogramming factors, Oct, Sox2, cMyc, and Klf4 induce an individual's somatic cells to revert to a more embryonic state (forming iPSC).
    (c) Induced pluripotent stem cells form unipotent stem cells which give rise to a specific cell type.

5. (a) In general, efficiency of conversion to iPSC is very low.
    (b) Some of the reprogramming factors used to form iPSC are proto-oncogenes and are implicated in cancer.

162. Using Stem Cells to Treat Disease (page 222)

1. The stem cells from the donor may not be immunologically compatible and may be rejected by the recipient's immune system.

2. If stem cells carrying a defective gene are placed into the patient, there will be no benefit. The disease will not be corrected because the stem cells carry the same genetic defect. The gene must be corrected.

3. If the disease is the result of a genetic defect, the defective gene will also be present in the umbilical cord blood. If the genetic defect is uncorrected, the treatment will not be effective.

4. (a) Stem cells can be cultured to develop into retinal pigment epithelium cells, which can be injected into the eye to replace damaged cells.
    (b) The patient's cells will carry the defect which will reappear in the new cells, so the treatment will not be effective.
    (c) Using a patient's own cells means that there is no chance of immune rejection of the tissue. This is useful when replacing an organ damaged by a non-genetic disease.

5. (a) Type 1 diabetes results from the body's immune system destroying the insulin-producing cells of the pancreas so that no insulin is produced and, as a consequence, the cells cannot take up glucose.
    (b) Stem cells taken from donors who do not have diabetes are induced to develop into insulin producing cells. They are then transplanted into the patient.

163. Using Totipotent Cells for Tissue Culture (page 224)

1. Micropropagation is a method used to produce plant clones, e.g. for the rapid multiplication of valuable varieties.

2. (a) A callus is an undifferentiated mass of cells.
    (b) The sterilised callus can be stimulated to produce roots and shoots by adding plant hormones to the culture medium in the appropriate concentrations.

3. Continued culture of a limited number of cloned varieties reduces genetic diversity and plants may become susceptible to disease and lack the ability to respond to environmental change.

4. Any two of:
    -   It is possible to create large numbers of clones from a single seed or explant.
    -   It is possible to select for desirable traits directly from the culturing setup, decreasing the amount of space required for field trials.
    -   Plants can be reproduced without having to wait for them to produce seeds.
    -   Rapid propagation is possible for species that have long generation times, low levels of seed production, or seeds that do not readily germinate.
    -   Enables the preservation of pollen and cell collections from which plants may be propagated.
    -   Allows the international exchange of sterilised plant materials (eliminating the need for quarantine).
    -   Helps eliminate plant diseases through careful stock selection and sterile techniques during propagation.
    -   It overcomes seasonal restrictions for germination.
    -   Enables cold storage of large numbers of viable plants in a small place.

5. (a) The stalked florets are meristematic and, being totipotent, can develop into the entire plant given an appropriate growth environment.
    (b) To sterilise them and remove any contaminants such as

fungi or bacteria.

(c) Aseptic techniques prevent the growth of unwanted contaminants such as fungi or bacteria.

(d) If one of the explants becomes mouldy during the cloning process, you could suspect that the floret had not been properly and completely sterilised.

164. Controlling Gene Expression: An Overview
(page 226)

1. DNA methylation, histone modification so the DNA is packed more tightly (e.g. addition of methyl groups to histones), mRNA degradation.

2. Modification of histones so that the DNA packs less tightly. Post transcriptional changes such as poly A tails and 5' caps slow mRNA degradation and extend the time mRNA is available for transcription.

3. DNA methylation and histone modification are ways in which gene expression may be altered. The level of histone modification and DNA methylation is influenced by the environment that an individual experiences. Over time, changes may accumulate that alter gene expression and therefore the individual phenotype.

4. The poly(A) tail and the guanine cap slow degradation of the mRNA by enzymes.

5. Exons can be spliced in different ways. They can be spliced in different arrangements or some exons can be left out and not appear in the mature mRNA. These differences create different versions of the final mRNA molecule and therefore different versions of the mRNA product (different proteins).

6. Post-translational changes include cleavage and addition of functional groups. Cleaving (cutting) the chain, is sometimes required to release the functional protein (e.g. insulin). Adding carbohydrate or lipid tags adds functional groups that alter the polypeptide's shape, final cellular destination, and function.

7. The large number of proteins formed from small numbers of genes is a result of modifications to the mRNA and the polypeptide chains. Post-transcriptional changes, such as exon splicing or addition of poly(A) tails, produce variations in the mRNA or its level of activity. Post-translational changes alter the shape or functional characteristics of the polypeptide. All these alterations collectively result in a large variety of proteins.

8. Protein degradation prevents their activity when they are no longer required and allows components to be recycled.

165. DNA Packaging and Control of Gene Expression
(page 228)

1. (a) DNA: A long, complex nucleic acid molecule found in the chromosomes of nearly all organisms (some viruses have RNA instead). Provides the genetic instructions (genes) for the production of proteins and other gene products (e.g. RNAs).

(b) Chromatin: Chromosomal material consisting of DNA, RNA, and histone and non-histone proteins. The term is used in reference to chromosomes in the non-condensed state.

(c) Histone: Simple proteins that bind to DNA and help it to coil up during cell division. Histones are also involved in regulating DNA function.

(d) Nucleosome: The basic unit of DNA packing consisting of a length of DNA wrapped around a bead of 8 histones.

2 (a) DNA methylation and histone modification can make DNA tightly pack together or remain loose depending on the type and placement of modification.

(b) Tightly packed DNA cannot be transcribed while loosely packed DNA can be transcribed.

166. DNA Methylation and Gene Expression (page 230)

1. (a) Genomic imprinting is a phenomenon in which the pattern of gene expression is different depending on whether the gene comes from the mother or the father.

(b) The imprinted genes are silenced by methylation and histone modification.

(c) Activity of the gene (including mutated versions) is determined by whether it is inherited from the mother or the father. The difference in expression may result in phenotype differences or genetic disorders.

2. The mother must have donated the imprinted (silenced) gene because if the father's gene was working correctly there would be no syndrome as there would still be a working copy of the gene. If we know the father's gene was mutated and the syndrome appears, then the mother's corresponding gene must be silenced.

167. Epigenetic Factors and Phenotype (page 231)

1.

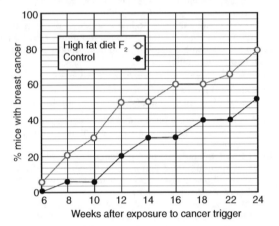

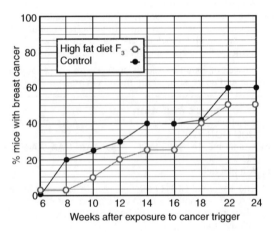

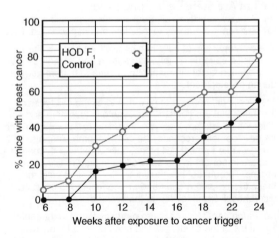

© 2016 **BIOZONE** International
ISBN: 978-1-927309-22-3
Photocopying Prohibited

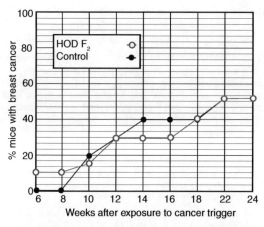

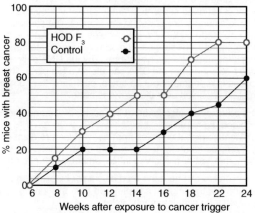

2. (a) F_1 and F_2
   (b) F_1, F_2, and F_3
   (c) The high oestrogen diet.

3. Experiments show that the effects of a parent's diet can be passed on to many following generations. Because the changes last beyond the F_2 generation these is evidence that the epigenetic effect is inherited.

168. The Role of Transcription Factors (page 233)
1. (a) Promoter: A DNA sequence where RNA polymerase binds and starts transcription.
   (b) Transcription factors: These are proteins that recognise and bind to the promoter sequence and to the enhancer sequence and thereby facilitate initiation of transcription.
   (c) Enhancer sequence: The DNA sequence to which the transcription factors called activators bind. This binding is important in bringing the activators in contact with the transcription factors bound to the RNA polymerase at the promoter.
   (d) RNA polymerase: The enzyme that, with the initial aid of transcription factors, transcribes the gene.
   (e) Terminator sequence: Nucleotide sequences at the end of a gene that function to stop transcription.

2. Condensing the chromatin is a way to regulate transcription. Chromatin in a condensed state cannot be transcribed because the transcription proteins cannot reach the DNA and the genes there cannot be expressed.

169. Oestrogen, Transcription, and Cancer (page 234)
1. Oestrogen binds to oestrogen receptors and acts as a transcription factor for the transcription of the AID gene.

2. AID causes somatic hypermutation in the DNA of B cells. This causes them to produce novel copies of antibodies which are produced against a myriad (as yet not encountered) antigens.

3. High levels of oestrogen allow for the greater production of AID because more oestrogen is available to initiate the transcription pathway. Normally, levels of oestrogen fluctuate in the body. In men they are relatively lower, but in women the

levels rise just before ovulation. There is a corresponding rise in immunity at this time.

4. Continual high levels of oestrogen result in the AID gene being switched on more than is normal, leading to a higher probability of a translocation mutation. This may also explain the link between women on the oral contraceptive pill and their slightly increased risk of cancer. AID's normal role is to cause role-driven (purposed) somatic hypermutation in B cells, but it can also cause other mutations. Specifically, it can cause a translocation mutation (of the CD95/Fas gene, a death receptor gene), preventing normal cell death and inducing tumour formation.

170. RNA Interference Models (page 235)
1. siRNA is formed from exogenous dsRNA while miRNA is formed from hairpin loops of non-protein-coding RNA copied from endogenous DNA.

2. RNAi regulates gene expression by cleaving mRNA molecules, thus rendering them inactive. In this way proteins that would be coded for by genes are not expressed as their protein product is not made.

3. MicroRNAs are involved in silencing genes and regulating gene expression. If microRNAs are under-expressed there will be poor/insufficient regulation of cells proliferating out of control (tumour cells).

171. Oncogenes and Cancer (page 236)
1. (a) Proto-oncogenes control the start of cell division. Tumour suppressor genes switch off cell division. The action of these two genes therefore regulates the timing of cell division.
   (b) Oncogenes remain switched on an continually signal a cell to divide. More cells than are needed are produced and a tumour forms.

2. Carcinogens are mutagenic agents capable of causing cancer.

3. Changes to cell chemistry in a primary tumour encourage the formation of capillaries, which provide a route (to other blood vessels) for malignant cells to break away from the primary tumour and spread throughout the body.

172. Oestrogen and Cancer (page 237)
1. Oestrogen is a steroid hormone responsible for normal primary and secondary sexual development in women and regulation of the menstrual cycle.

2. High levels of oestrogen are associated with an increased risk of cancer, particularly breast cancer.

3. The risk of breast cancer increases with age, up to age 64 where the risk levels off (plateaus).

4. (a) Having children at a later age increases a woman's risk of breast cancer.
   (b) Having children at a younger age reduces lifetime exposure to oestrogen because pregnancy and breast feeding suppress oestrogen levels.

5. Contraceptive pills increase the risk of developing breast cancer.

173. Genome Projects (page 238)
1. (a) One of: May lead to improvements in milk/meat quality. May tell us more about the impact of domestication on the genetics of a once wild species. By comparing beef and dairy breeds, it may also aid understanding of the mechanisms of gene expression.
   (b) A model organism. Allows geneticists to identify genes with similar functions in humans.
   (c) Immune system is complex so could help in studying human immune function and immune disorders.

2. The cost and time to sequence a human genome equivalent have decreased exponentially from one billion dollars to three thousand dollars in ten years.

3. 9317

© 2016 **BIOZONE** International
ISBN: 978-1-927309-22-3
Photocopying Prohibited

174. Genomes, Bioinformatics, and Medicine
(page 239)
1. A genome can be sequenced and added to the database of existing genomes. Genes are then identified, cross referenced with other known similar genes, and their functions investigated. Existing drugs targeting similar genes or gene products can be identified and their effectiveness against the newly identified genes and their products tested. Once a gene product is known, it can be synthesised and tested on a model organism so that new vaccines or drugs can be developed to target the newly identified genes.

2. The completion of the *P. falciparum* genome has enabled the identification of specific gene loci that encode antigens. These can be used to develop a vaccine against malaria (antigenic material is the basis of vaccination). However, few of the loci are shared by all *Plasmodium* species, so a single vaccine against all species is unlikely.

175. The Human Genome Project (page 240)
1. The HGP aimed to map the entire base sequence of every chromosome in the human cell (the human genome) and identify and map the genes.

2. (a) 53%
   (b) 8%
   (c) Knowing the DNA sequence and position of genes will enable better screening for genetic diseases and the development of drugs or other treatments for them.

3. Chromosome 9

176. Making a Synthetic Gene (page 241)
1. (a) Synthetic DNA is constructed in the 3' to 5' direction, opposite to natural DNA synthesis.
   (b) Oligonucleotides are used to prevent the DNA bases incorrectly bonding to each other (the longer the chain the more chance of incorrect DNA bonding).

2. New genes that do not exist in nature may be produced. These may be used to produce new materials for industry or new therapeutic drugs.

177. Making a Gene from mRNA (page 242)
1. Restriction enzymes are used to produce the DNA fragment (often a human gene) that is to be cloned, by isolating it from other DNA and providing it with sticky ends. The same enzymes are used to open up the plasmid or viral DNA into which the DNA fragment is to be inserted.

2. (a) Introns are non-protein coding and removing the makes the gene shorter and therefore easier to insert into vectors and easier for bacteria to translate into the protein produce. In the case of PCR, it means large amounts of non-coding DNA are not made.
   (b) Reverse transciptase catalyses the production of a DNA strand from a strand of RNA.

3. Reverse transcriptase is found in retroviruses (such as HIV) and make a copy of DNA from the viral RNA so that the viral genes can be integrated into (and replicated along with) the host's genome.

178. Making Recombinant DNA (page 243)
1. Restriction enzymes cut DNA into lengths or to isolate genes by cutting at specific recognition sites.

2. Sticky ends on DNA allow different DNA strands cut with the same restriction enzyme to be joined (via the complementary overhanging base pairs. Blunt ends on DNA strands allow DNA strands of any other blunt end fragments to be joined. The strands do not have to be complementary.

3. Having many different kinds of restriction enzymes allows DNA to be cut at many different recognition sites and so produce a variety of sticky or blunt ends. This allows for a better ability to isolate and join different regions of the DNA.

4. (a) The two single-stranded DNA molecules are recombined into a double-stranded molecule. This is achieved by H-bonding between complementary bases.

   (b) DNA ligase joins two adjacent pieces of DNA by linking nucleotides in the sticky ends.

5. It joins together DNA molecules, while restriction digestion (by restriction enzymes) cuts them up.

6. With a few exceptions, all organisms on Earth use the same DNA code to store information and the same cellular machinery to read the information and express it. Any DNA from any organism can therefore be read and expressed by the cellular machinery of any other organism into which the DNA is spliced.

179. DNA Amplification Using PCR (page 245)
1. To produce large quantities of 'cloned' DNA from very small samples. Large quantities are needed for effective analysis. Very small quantities are often unusable.

2. Detail not necessarily required in the answer bracketed. A double stranded DNA is heated (to 98°C for 5 min), causing the two strands to separate. Primers, free nucleotides, and DNA polymerase are added to the sample. The sample is then cooled (to 60°C for a few minutes), and the primers anneal to the DNA strands. The sample is incubated and complementary strands are created (by the DNA polymerase) using each strand of the DNA sample as a template. The process is repeated about 25 times, each time the number of templates doubles over the previous cycle.

3. (a) Forensic samples taken at the scene of a crime (e.g. hair, blood, semen).
   (b) Archaeological samples from early human remains.
   (c) Samples taken from the remains of prehistoric organisms mummified or preserved in ice, amber, or tar pits etc.

4. This exercise can be done on a calculator by pressing the 1 button (for the original sample) and then multiplying by 2 repeatedly (to simulate each cycle).
   (a) 1024   (b) 33 554 432 (33.5 million)

5. (a) It would be amplified along with the intended DNA sample, contaminating the sample and rendering it unusable.
   (b) Sources of contamination (any two of):
       Dirty equipment (equipment that has DNA molecules left on it from previous treatments).
       DNA from the technician (dandruff from the technician is a major source of contamination!)
       Spores, viruses and bacteria in the air.
   (c) Precautions to avoid contamination (any two of):
       Using disposable equipment (pipette tips, gloves).
       Wearing a **head cover** (disposable cap).
       Use of **sterile procedures**.
       Use of **plastic disposable tubes with caps** that seal the contents from air contamination.

6. (a) and (b) any of the following procedures require a certain minimum quantity of DNA in order to be useful: DNA sequencing, gene cloning, DNA profiling, transformation, making artificial genes. Descriptions of these procedures are provided in the workbook.

180. In Vivo Gene Cloning (page 247)
1. *In vivo* methods make it simpler to produce the protein product because the recipient of the gene (the bacterium) will express the gene as its protein product and this product can then be harvested.

2. One would not use bacteria to clone a gene if the purpose of the gene cloning was simply to amplify the DNA, e.g. for forensic or diagnostic purposes, and protein expression was not required. PCR alone is a simple automated process.

3. The gene is prepared by removal of introns. At the same time, an appropriate vector (e.g. plasmid) is isolated. Both the gene and plasmid are treated with the same restriction enzyme to produce identical sticky ends. The DNA fragments are mixed in the presence of DNA ligase and anneal (DNA ligation).

4. 2 x 24 = 48 replications in twenty four hours. Number of bacteria and therefore plasmids = 2^{48} or 2 multiplied by itself 48 times = 281,474,976,710,656 copies.

5. Recombinant colonies can be identified by their ability to grow

on agar with ampicillin but not tetracycline.
i) Grow the bacteria on agar containing ampicillin. All resulting colonies must contain the plasmid.
ii) Press a sterile filter paper firmly onto the surface of the agar, taking care to mark the paper's position relative to the agar plate. Press the paper onto another agar plate containing tetracycline and mark the position of the paper relative to the plate. Any colonies that do not grow on this new agar must contain the recombinant DNA (as they have been killed by tetracycline).
iii) Match the position of colonies between the first and second agar plates. Those on the first plate that are missing from the second are isolated and cultured.

6. A *gfp* marker is preferable over antibiotic resistance genes because antibiotic resistance genes encourage the undesirable spread of antibiotic resistance through bacterial populations. Apart from this, a *gfp* marker system is much simpler, as the marked colonies can be quickly identified by fluorescence under UV light.

181. Making Chymosin Using Recombinant Bacteria

(page 249)
1. Chymosin (an enzyme) is used to coagulate milk into curds in the production of cheese.

2. Traditional source of chymosin was from the stomachs of young (suckling) calves.

3. (a) Restriction enzymes can be used to cut DNA at specific sites and joined again with DNA ligase. Genes of interest can be isolated and inserted into vector DNA e.g. plasmid.
   (b) Bacteria can take up modified plasmid DNA and replicate it in culture.
   (c) The amino acid sequence of a gene protein product and its mRNA sequence can be determined.
   (d) Reverse transcriptase can synthesise DNA from mRNA to construct a protein-coding gene.
   (e) Large quantities of the protein can be produced by growing the bacteria in large batches.

4. Chymosin gene was isolated by first determining its amino acid sequence and from that, the mRNA coding sequence. Once this is known, a gene probe can be constructed and used to locate the mRNA of interest. Reverse transcriptase is used to produce DNA strand, which can then be amplified using PCR. This technique can be applied to the isolation of any gene of interest, once its protein product is known.

5. Advantages of using GE chymosin, (a)-(c) any 3 of:
   – The chymosin produced is identical to chymosin from its natural source.
   – The chymosin can be produced without the unnecessary slaughter of calves (a welfare issue for many people) and is suitable for vegetarians.
   – The chymosin is 80-90% active ingredient and is thereby significantly purer than natural rennet, which contains only between 4-8% active chymosin.
   – The chymosin can be produced on demand, in the quantities required. This makes it cost effective.

6. Fungi are eukaryotes. The cells are larger and their secretory pathways are more similar to those of humans than those of *E.coli.*

182. Insulin Production (page 251)
1. (a) High cost (extraction from dead tissue is expensive). Non-human insulin is different enough from human insulin to cause side effects. The extraction methods did not produce pure insulin so the insulin was often contaminated.
   (b) Using recombinant DNA technology to produce human insulin provides a way to produce a low cost, reliable supply for consumer use. The insulin protein is free of contaminants and, because it is a human protein, the side effects of its use are minimised.

2. The insulin is synthesised as two (A and B) nucleotide sequences (corresponding to the two polypeptide chains) because a single sequence is too large to be inserted into the bacterial plasmid. Two shorter sequences are small enough to

be inserted (separately) into bacterial plasmids.

3. The β-galactosidase gene in *E.coli* has a promoter region so the synthetic genes must be tied to that gene in order to be transcribed.

4. (a) Insertion of the gene: The yeast plasmid is larger and can accommodate the entire synthetic nucleotide sequence for the A and B chains as one uninterrupted sequence.
   (b) Secretion and purification: Yeast, a eukaryote, has secretory pathways that are more similar to humans than those of a prokaryote and the β-galactosidase promoter is not required so secretion of the precursor insulin molecules is less problematic. Purification is simplified because removal of β-galactosidase is not required and the separate protein chains do not need to be joined.
   (c) Yeast would be most preferred because it can express both parts of the insulin protein as an interrupted sequence so the process of refining the protein is simpler.

183. Genetically Modified Food Plants (page 253)
1. The genes for two different enzymes involved in beta carotene synthesis are taken from two different sources and inserted into the nuclear genome of a rice plant. Expression of the gene under the control of an endosperm specific promoter results in production of beta carotene in the edible portion of the rice plant.

2. The expression of the genes is controlled by a promoter specific to the endosperm, so the genes will only be expressed in that tissue.

3. *Agrobacterium tumefaciens* is a natural plant pathogen and can transfer genes as a consequence of infecting a host plant. The tumour-inducing Ti plasmid can be modified to delete the tumour-forming gene and insert a gene for a desirable trait.

4. (a) Beta carotene is a vitamin A precursor. Production and consumption of beta-carotene rich rice could alleviate or prevent diseases related to vitamin A deficiency (e.g. night blindness).
   (b) Improved nutrition through GM rice will be viable only if the diet in targeted regions is also adequate with respect to fat intake. In some impoverished regions this will not be the case, as diet is inadequate across a wide range of food groups, including fat and protein.

5. (a) More grains per head, larger grains.
   (b) Faster maturation time.
   (c) Improved resistance to pests (e.g. by producing a natural toxin or thicker seed coat etc).

184. Gene Therapy (page 255)
1. (a) To correct a genetic disorder of metabolism by correction, replacement, or supplementation of a faulty gene with a corrected version.
   (b) Gene therapy might be used for inherited genetic disorders of metabolism, non-infectious acquired diseases (e.g. cancer), and infectious diseases.

2. Transfection of germline cells allows the genetic changes to be inherited. In this way, a heritable disorder can be corrected so that future generations will not carry the faulty gene(s). Transfection of somatic cells only corrects those cells for their lifetime.

3. Gene amplification is used to make multiple copies of the normal (corrective) allele.

4. GM stem cells offer longer lasting treatment because the cells will continue proliferating and the therapeutic role will continue as long as the cell line continues. Once a transformed (corrected) somatic cell has reached the end of its life, its therapeutic role ends.

5. (a) Viruses are good vectors because they are adapted to gain entry into a host's cells and integrate their DNA into that of the host.
   (b) Problems include (two of):
   • Host can develop a strong immune response to the viral infection. In immune-suppressed patients, this could severely undermine their health.
   • Retroviruses infect only dividing cells.

- Viruses may be eliminated by host's immune system.
- If they do not integrate into the chromosome, the inserted genes may only function sporadically.
- Genes may integrate randomly into chromosomes and disrupt the functioning of normal genes.

6. (a) If a therapeutic gene is integrated into the chromosome, it has a better chance of normal long term function and stability.
   (b) When it integrates into the host's (patient's) chromosome, the gene has the potential to disrupt normally functioning genes.

7. (a) Naked DNA is unstable because it is recognised as foreign and is easily degraded by the normal defence mechanisms of the host. Uptake by cells is inefficient and, once within the cell, the DNA is still at risk of degradation by lysosomes.
   (b) Liposomes offer greater stability because they are formulated to be recognised and targeted by the host cell's receptors. They are therefore less likely to be degraded in the tissues.

---

185. Gene Delivery Systems (page 257)

1. (a) CF symptoms include disruption of the function of secretory glands including the pancreas, intestinal glands, biliary tree, sweat glands, and bronchial glands. Infertility. Disruption of lung function caused by an accumulation of thick, sticky mucus in the lungs.
   (b) CF has been targeted because the majority of cases are the result of a gene defect involving the loss of only one triplet (three nucleotides). In theory, correction of this one gene should not be difficult.
   (c) Correction rate has been low (25%), and the effects of correction have been short lived and the benefits quickly reversed. These problems are related to the poor survival of the viral vector in the body and the sporadic functioning of the gene because it is not integrated into the host's (human) chromosome. Patients may also have an immune reaction to the vector. In one patient, treatment was fatal.

2. (a) **Vector**: Adenoviruses.
      Potential problems (not required): Poor integration and survival of vector. Immune reactions. Low corrective rates.
   (b) **Vector**: Liposomes
      Potential problems (not required): Liposomes are less efficient than viruses at transferring genes, so corrective rates are lower than for viral vectors.

3. (a) X-linked SCID is caused by a mutation to a gene on the X chromosome that encodes for the common gamma chain. ADA SCID is caused by a defective gene that codes for adenosine deaminase.
   (b) The vector used in the treatment of SCID is a gutted retrovirus (a retrovirus with its natural genetic material removed to make room for the corrective gene being transferred).

4. Gene therapy for cystic fibrosis is targeted at the lungs using either an adenovirus or a liposome delivered via the airways. SCID is treated using a retrovirus, which is introduced to the patient's own bone marrow stem cells. These are then cultured and returned to the patient.

5. (a) Chance of interfering with essential gene function: When an essential gene function is affected by gene therapy in somatic cells, the individual will be affected. There may be a chance of corrective therapy in that person's lifetime. When the change affects germline cells, all descendants of the treated individual have a chance to inherit the disrupted gene, so a second heritable defect is created.
   (b) Misuse of the therapy to selectively alter phenotype: Alteration of somatic cells to selectively alter one's phenotype is (presumably) a matter of one's own choice; it may benefit that person in their own lifetime, but will not affect subsequent generations. When these selective changes affect the germline cells, then they are heritable and the alteration is not necessarily limited to one individual. This poses the problem of genetic selection and eugenics, and all their consequent ethical dilemmas.

---

186. The Ethics of GMO Technology (page 259)

1. (a) Advantage: Crop growers would not need to spray for crops pest as often.
   (b) Problem: Plants producing toxins as a pest resistance may cause health problems if eaten over a long period of time. Pest resistant plants could become a problem outside the crop field (as weeds).

2. The widespread use of antibiotic marker genes in food crops for human consumption or stock food may give rise to antibiotic resistant strains of pathogenic bacteria which affect humans and stock animals. Restrained use of antibiotics is now considered essential in preventing large scale development of antibiotic resistance.

---

187. Food for the Masses (page 260)

1. The bacterium because it uses only one enzyme to facilitate multiple reactions. It is therefore simpler to use in the production of the modified plant.

2. The gene can be isolated by first identifying the associated enzyme and its amino acid sequence. From this, the mRNA sequence can be identified. The correct mRNA molecules can then be extracted and reverse transcriptase used to copy the mRNA into DNA. This can then be amplified.

   Or - Once the DNA sequence is identified, PCR primers can be produced that will anneal to the start sequences of the gene. The PCR product will be the targeted gene.

   Or - The gene can be identified from its protein product (as above). This can then be cut from the chromosome using restriction enzymes and amplified.

3. *Agrobacterium*. This bacterium can transfer DNA into plants, so that the cells will contain recombinant DNA.

4. A plasmid (the *Ti plasmid*) is removed from the *Agrobacterium*. Using restriction enzymes the plasmid is cut and the target DNA inserted (the tumor forming gene is removed). DNA ligase is used to attach the target DNA to the plasmid. The plasmid is then replaced into the *Agrobacterium*.

5. (a) *Agrobacterium* transfers the recombinant plasmid to the cells of the target plant.
   (b) The best stage of development for transformation is while the plant is in embryonic form. In this way a larger proportion of the plant cells will be affected and take up the new DNA. This will cause a much better result in the adult plant.

6. Transformed plants can be identified if an extra gene is inserted along with the target gene. This is normally a gene for drug resistance. Plants grown on agar impregnated with the chemical will grow only if they have taken up the new DNA. Those without won't grow.

7. A large number of plants can be produced using plant tissue culture or vegetative propagation. In this way many plants can be produced, which will lead to rapid dissemination of the transgenic stock.

---

188. Gel Electrophoresis (page 262)

1. Purpose: To separate mixtures of molecules (proteins, nucleic acids) on the basis of size, electric charge and other physical properties.

2. (a) The frictional (retarding) force of each fragment's size (larger fragments travel more slowly than smaller ones).
   (b) The strength of the electric field (movement is more rapid in a stronger field). Note: The temperature and ionic strength of the buffer can be varied to optimise separation of the fragments.

3. The gel is full of pores (holes) through which the fragments must pass. Smaller fragments pass through these pores more easily than larger ones.

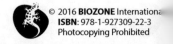

189. Interpreting Electrophoresis Gels (page 263)

1. (a) and (b)

Cow synthesised DNA:
TGA TTG TAA GCT TTC AGG GTG GGT GAT TA

Cow sample DNA:
ACT AAC ATT CGA AAG TCC CAC CCA CTA AT

Sheep synthesised DNA:
TAG TTG TAG GCT TTT TGG GTG GGT GAT TA

Sheep sample DNA:
ATC AAC ATC CGA AAA ACC CAC CCA CTA AT

Goat synthesised DNA:
TGG TTG TAG GCT TTC TGG GTG GGT AAT TA

Goat sample DNA:
ACC AAC ATC CGA AAG ACC CAC CCA TTA AT

Horse synthesised DNA:
TGT TTG TAG GCC TTT AGA GTG GGT GAT TA

Horse sample DNA:
ACA AAC ATC CGG AAA TCT CAC CCA CTA AT

(c) Sheep and goat (3 differences)
(d) Goat and horse (6 differences)

2.

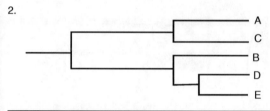

190. Screening for Genes (page 264)

1. A DNA probe allows a region of DNA to be marked and identified.

2. The DNA probe is a sequence of DNA that is complementary to the target region of DNA. It is therefore able to bond to the target DNA allowing that region to be identified.

3. Double stranded DNA must be denatured to a single strand so that the base pairs are exposed and the complementary base pairs on the DNA probe can bond to the target region.

4. Genes are visualised using fluorescent light or X-ray film, depending on the tag that has been added to the probe being used. The target DNA sequence will appear as a band on the electrophoresis gel.

191. Hunting for a Gene (page 265)

1. The physical effects of Huntington's disease are shaking of hands and/or limbs and an awkward gait. More severe effects include the loss of muscle control and mental function leading to dementia.

2. The mHTT gene was discovered using information from the family history of 10,000 people. Using a probe called G8, a map of the 4th chromosome was built up and each gene sequenced. The mHTT gene was shown to be one with a trinucleotide repeat expansion.

3. HD is caused by a trinucleotide repeat expansion of the sequence CAG on the 4th chromosome. Repeats of over 35 cause the disease and the greater the number of repeats the more severe the disease. Because of the instability of the mHTT gene the number of repeats and severity of the disease tends to increase over generations.

192. Genetic Screening and Embryo Selection (page 266)

1. The main purpose of PGD is to identify embryos that may have genetic abnormalities before they are implanted.

2. PGD uses genetic screening techniques to identify genetic defects in an embryo when one or both of the parents are known to carry a genetic abnormality. One cell from an embryo is extracted and the DNA screened for genetic disease. Embryos are produced by normal fertility treatments and only the embryos without genetic abnormalities are implanted in the mother's uterus.

3. The polar body is extracted from the unfertilised egg. This will contain only the maternal half of the embryo's complete DNA. Therefore, screening can only detect those genetic diseases carried by the mother.

4. PCR amplifies small amounts of DNA. Because the template DNA is such a small amount, it is sensitive to contamination and thus could produce a false response by amplifying contaminant DNA.

5. PGD can be used to screen for genetic markers in embryos so that genetically desirable livestock embryos can be identified and only these isolated and implanted. This pre-selection saves time and money because undesirable stock are eliminated are do not need to be raised as stock.

6. Ethical issues in PGD usually arise over discrimination issues and destruction of embryos. Points may include:
   - Some may protest over the destruction of an embryo which they see as life. However, many would point out that life with a debilitating genetic disease is not the quality of life many would want.
   - There is also the option of aborting pregnancies that are found to have genetic defects. Is this any different to destroying the embryo when it is only eight cells?
   - There is also the issue over discrimination. If we select embryos without a genetic disease, are we saying that people with that genetic disease are somehow not as entitled to life as those without it?
   - PGD provides the ability to select the sex of the embryo. Many people may say this should not be allowed, but if a family already has (for example) 3 girls, should they be allowed to ensure the next child is a boy?
   - Being able to screen for sex and genetic traits could be regarded as the beginning of designer babies. Should parents be able to chose the traits of their child and is this a disservice to the child?

193. DNA Profiling Using PCR (page 268)

1. STRs (microsatellites) are non-coding nucleotide sequences (2-6 base pairs long) that repeat themselves many times over (repeats of up to 100X). The human genome has numerous different STRs; equivalent sequences in different people vary considerably in the numbers of the repeating unit. This property can be used to identify the natural variation found in every person's DNA since every person will have a different combination of STRs of different repeat length, i.e. their own specific genetic profile.

2. (a) Gel electrophoresis: Used to separate the DNA fragments (STRs) according to size to create the fingerprint (profile).
   (b) PCR: Used to make many copies of the STRs. Only the STR sites are amplified by PCR because the primers used to initiate the PCR are very specific.

3. (a) Extract the DNA from sample. Treat the tissue with chemicals and enzymes to extract the DNA, which is then separated and purified.
   (b) Amplify the microsatellite using PCR. Primers are used to make large quantities of the STR.
   (c) Run the fragments through a gel to separate them. The resulting pattern represents the STR sizes for that individual (different from that of other people).

4. To ensure that the number of STR sites, when compared, will produce a profile that is effectively unique (different from just about every other individual). It provides a high degree of statistical confidence when a match occurs.

194. Forensic Applications of DNA Profiling (page 270)

1. Profiles of everyone involved must be completed to compare their DNA to any DNA found at the scene and therefore eliminate (or implicate) them as suspects.

2. The alleged offender is not guilty. The alleged offender's DNA profile does not appear in the DNA collected at the crime scene nor does it appear in the DNA database. Profile E's DNA is found at the scene.

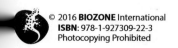
© 2016 **BIOZONE** International
ISBN: 978-1-927309-22-3
Photocopying Prohibited

3.   (a) 0.0294
    (b) 0.3446
    (c) 2pq = 2 x 0.0294 x 0.3446 = 0.0203

4.   14-10, 14-11, 15-10, 15-11

5.   (a) No
    (b) The man cannot be the biological father because there are two mismatches in the profiles. The child does not show any matches with STR D19S433 and D2S441.

6.   (a) Each whale species has a distinct DNA profile. Profiling the whale meat therefore reveals the types of whales they meat came from.
    (b) Individual whales also have their own DNA profile. Profiling the whale meat from each species reveals how many whales of that species were killed (by simply counting the number of different profiles.

195. Profiling for Analysis of Disease Risk (page 272)
1.   (a) Haplotypes are used because the many SNPs in them fall into distinct combinations that can be matched statistically to a particular condition. Single SNPs will not provide enough information; there would be no way to match a particular SNP to a particular condition.
    (b) In most cases, there will be people with the disease and people without the disease who have the same SNP or haplotype profile. Therefore the haplotype only gives a probability of a person having the disease.

2.   Genomic analyses can provide probabilities for successful treatment outcomes, i.e. for a given genetic profile, a particular treatment will have a certain probability of being successful. This helps match drugs to a person's personal genetic profile based on their sequence variations.

196. Finding the Connection (page 273)
1.   Microsatellite DNA sequences will be used.

2.   The DNA primers are used as starting points for the PCR process. They enclose the section of DNA that will be isolated and amplified.

3.   Primers are annealed to the beginning of the target sequence of DNA. DNA (Taq) polymerase then copies the target sequence. After many cycles a large number of copies of the target DNA are produced.

4.   Electrophoresis is used to separate the lengths of microsatellite DNA. Different lengths of DNA will appear as bands in different places on the electrophoresis gel.

5.   Profiles that have similar banding indicate relatedness in the breeds. The more disparate the banding the less related breeds are. Banding patterns can be computer analysed and a dendrogram produced.

197. Chapter Review (page 274)
    There is no model answer for this activity. The summary is the student's own.

198. Key Terms: Did You Get It? (page 276)
1.   Adult stem cell (P), annealing (B), DNA amplification (J), DNA ligation (L), DNA polymerase (O), embryonic stem cells (Q), gel electrophoresis (I), GMO (N), marker gene (G), microsatellite (M), mutation (R), PCR (H), primer (D), recognition site (F), recombinant DNA (K), restriction enzyme (E), sticky end (C), vector (A).

2.   (a) Base deletion (T)
    (b) Base insertion (A)
    (c) Base substitution (G for T)
    (d) Inversion GGC GCT →TCG CGG
    (e) Triplet deletion (TTT)

3.   (a)

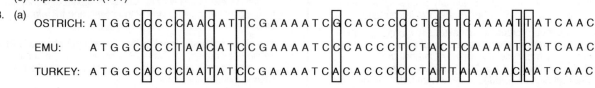

    (b) There are 5 differences between the ostrich and emu.
    (c) There are 9 differences between the ostrich and turkey.
    (d) The ostrich and emu are most closely related.
    (e) The ostrich and emu have the fewest differences in their gene sequence.
    (f) Yes, the sequence information supports similarities based on appearance. The emu and ostrich both have similar body proportions with long legs and necks, long shaggy feathers, and a similar foot structure with reduced number of toes (two in ostriches and three in emus).

4.   (a) Cardiomyocytes and other precursor cells that can differentiate only along one lineage, such as hepatoblasts (liver stem cells).
    (b) Hematopoietic stem cells (i.e. bone marrow stem cells and cells in umbilical cord blood), mesenchymal stem cells (e.g. bone stem cells and adipose stem cells), epithelial stem cells (most epithelia comprise a number of cell types although some epithelial stem cells may be unipotent).
    (c) Embryonic stem cells
    (d) The zygote and its first few divisions.

5.

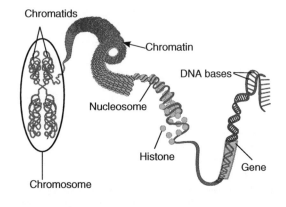

© 2016 **BIOZONE** International
ISBN: 978-1-927309-22-3
Photocopying Prohibited